Smith Ely Jelliffe, Henry Hurd Rusby

Essentials of Vegetable Pharmacognosy

A treatise on structural botany. Designed especially for pharmaceutical and medical

students, pharmacists and physicians.

Smith Ely Jelliffe, Henry Hurd Rusby

Essentials of Vegetable Pharmacognosy
A treatise on structural botany. Designed especially for pharmaceutical and medical students, pharmacists and physicians.

ISBN/EAN: 9783337370565

Printed in Europe, USA, Canada, Australia, Japan

Cover: Foto ©berggeist007 / pixelio.de

More available books at **www.hansebooks.com**

ESSENTIALS

OF

VEGETABLE PHARMACOGNOSY

A Treatise on Structural Botany.—Designed especially for Pharmaceutical
and Medical Students, Pharmacists and Physicians.

PART I.

THE GROSS STRUCTURE OF PLANTS.

By HENRY H. RUSBY, M. D.,

Professor of Botany, Physiology and Materia Medica in the College
of Pharmacy of the City of New York.

PART II.

THE MINUTE STRUCTURE OF PLANTS.

By SMITH ELY JELLIFFE, M. D.,

Professor of Pharmacognosy in the College of Pharmacy
of the City of New York.

WITH 560 ILLUSTRATIONS.

Mo. Bot. Garden,
1896.

NEW YORK.

D. O. HAYNES & CO.

1895.

PREFACE.

THE contribution of the matter here reprinted as a part of a series of articles on the pharmaceutical sciences in a weekly journal has necessitated a condensed and abbreviated mode of treatment. Nevertheless, it represents fairly well the teaching of this portion of the subject as followed in the author's class room, and intelligently displays the relations of morphology and organography to the study of vegetable drugs. The attempt has been made to allow richness of illustration to take the place of that full treatment in the text which has been precluded by the conditions under which the work was performed. Few of the illustrations have heretofore been used in botanical text books. Most of them have been taken from nature. A large number have been taken, by permission, from Engler & Prantl's Pflanzenfamilien and other sources, but specific acknowledgments in connection with the individual figures have not been possible in the serial form of the article.

TABLE OF CONTENTS.

	PAGE.
Introduction	1
Departments of Scientific Botany	2
Departments of Economic Botany	3
Pharmaceutical Botany	3
Order of Subjects	5
Anthology	5
The General Nature of the Flower	5
The Essential Organs and their Parts	10
The Non-Essential Organs	11
The Modifications of Floral Structure Classified	13
The Laws of Floral Structure, (number, position, form and size, separation and union of parts)	13
The Perigone	20
The Androecium	27
The Gynæcium	31
Forms of the Style	32
Forms of the Stigma	32
Structure and Forms of the Ovary	34
Positions of the Ovule	35
Structure and Forms of the Ovule	36
The Torus	37
The Disk	38
Pollination	39
Fertilization	43
Carpology	44
Changes Produced by Fertilization in Ovary and Ovule	44
Functions of the Fruit and their Performance	46
Structure of the Fruit	47
Classification of Fruits, with Key	51
Structure and Forms of the Seed	57
The Embryo	57
Germination	58

General Structure of Root and Stem.................................... 61

The Root, its Fundamental, Primary and Secondary States of Structure. 62

The Stem, the same.. 66

The Bark... 70

Classification of Roots.. 72

Classification of Stems.. 73

Modified Stems... 77

Rhizomes... 78

Buds and Bulbs... 78

General Structure of the Leaf.. 79

The Stipule.. 80

Classification of Leaves as to—

        Duration 81

        Texture.. 81

        Surface 82

        Attachment to the Plant.............................. 83

        Attachment of Blade to Petiole....................... 83

        Outline 83

        Apex 85

        Base 86

        Venation 87

        Margin... 87

        Division 89

Compound Leaves.. 90

Modified Leaves.. 93

Phyllotaxy 94

Antidromy 96

Anthotaxy.. 97

Flower Stems... 97

Involucres... 98

Classification of Inflorescences, with Key........................... 99

Special Characters of the Hypanthodium............................... 100

Special Characters of the Glumaceae.................................. 101

ESSENTIALS
OF
VEGETABLE PHARMACOGNOSY.

STRUCTURAL BOTANY.

INTRODUCTION.

Living Bodies differ from lifeless in their ability to grow by assimilating to their own substance extraneous and dissimilar substances, as seen in the use of carbonic acid in the production of starch and cellulose. This process is called Assimilation. They consist also of more or less distinct parts, each of which performs special work differing from that performed by the other parts. These parts are called Organs and the special work which each performs is called its Function. Living bodies are, therefore, designated as Organic Bodies and the part of nature composed of them the Organic Kingdom. Of the two terms that of "organic body" is usually preferable, as it applies equally well to a body in which life has ceased to exist. A third important characteristic of living bodies which may be mentioned is their power to give origin to other independent living bodies, which, separating from their parent, or remaining attached thereto, grow into a resemblance to it. That is, they possess the power of Reproduction.

The assimilated matter of organic bodies is called Organic Matter. Organic matter may be living, as cellulose, or lifeless, as starch. It may, as in the case of the starch, be prepared for future use as food, for the construction of tissue, as in the case of cellulose, or it may exist as disassimilated matter resulting from the performance of function. The latter may still be of some service in the plant economy, or, perhaps, be entirely useless. The energy and force of living matter are called Vital. The same laws govern the conservation of vital as of other force, though the manifestation is different and the working of the law is not so patent. There is a direct ratio between vital energy and the energy represented in the inorganic matter consumed, though the relations are complex and the estimations necessarily difficult, owing especially to the direct contribution of energy by the sun.

Organic bodies are of two kinds—Vegetable and Animal—and are denominated respectively, Plants and Animals. A third class was once recognized as consisting of those not distinctively either animal or vegetable. It is not now believed that any such exist, although no one means has yet been devised for ascertaining in every case to which sub-kingdom a body belongs. This fact, however, does not imply any difficulty in applying such distinctions in pharmacognosy, for all plants and animals which contribute to the materia medica are readily distinguishable. The only known invariable characteristic distinguishing plants from animals is the ability of the former, when not modified by parasiticism, to subsist exclusively on foods consisting of inorganic matter. Still, even independent plants are not entirely devoid of ability to consume organic matter with advantage, and animals require abundant supplies of inorganic matter as a portion of their food.

There are many other characteristics which distinguish those plants which are under common observation from animals, of which the following may be mentioned:

1.—The consumption of an excess of carbonic acid over that of oxygen and the discharge of an excess of oxygen as waste.

2.—The absence of a distinct digestive cavity.

3.—The lack of the power of locomotion. This term is not at all equivalent to that of "movement," for distinct and characteristic movements are present in all plants.

The study of the organic kingdom constitutes Biology. Biology in attention to the structure of bodies is Anatomy. Biology in attention to functions is Physiology. We have therefore both animal and plant anatomy, and animal and plant physiology. Biology relating to plants is Botany. Owing to the totally different methods of examination employed in the two cases it becomes of the greatest convenience to divide anatomy, in practice, into two parts. That part depending upon observations which can be pursued without the aid of the compound microscope is known as Gross Anatomy. That which requires such aid is Minute Anatomy, or Histology. Applied to botany the latter is commonly known as Microscopical Botany, a term which, though incongruous, possesses the excellent merit of being highly convenient and thoroughly expressive. As the study of botany involves the use of physics and chemistry, it is apparent that when so applied they become parts of botany, just as botany becomes a part of physics or chemistry when applied in the pursuit of those branches. The propriety of such terms as "chemical botany" or "botanical chemistry" is thus explained.

In accordance with fundamentally different considerations which concern us in studying botany at different times, various departments of the science are recognized. When such considerations have the attainment of scientific truth as their aim, these several departments pertain to Scientific Botany, as, for instance, a study of the natural processes of fertilization of the ovule. If upon the other hand the object be the utilization of such truths in advancing the arts, they belong to Economic or Applied Botany, as, for instance, the artificial fertilization of ovules in such a way that the resulting seeds will produce new and superior varieties of useful plants.

DEPARTMENTS OF SCIENTIFIC BOTANY.

The departments of scientific botany, and the manner in which one may arise from the necessities of another and contribute to it, may be illustrated as follows: It being understood that no plants are now in

existence which existed in the beginning, all having originated through changes effected in some manner in those which formerly existed, one of the great objects of botanical study is to ascertain the genetic relationships which exist between plants and to constitute such a systematic arrangement of them as shall, so far as practicable, indicate the lines and order of their development from others. This department constitutes Systematic Botany. Since such classification is based chiefly upon structure it is necessary that there should be a department known as Structural Botany. Before the facts ascertained by the structural botanist can be utilized in classification it is necessary that the relative ranks of the structural characteristics should be determined. Of any two structural characteristics that which was first developed, or is the older in creation, should form the basis of the primary division of the group, the other of a sub-division. In ascertaining such relative ranks a consideration of the uses of the several characters is of great value, so that Physiological Botany or Vegetable Physiology is brought into service. The genesis of plants is in no way more certainly indicated than in the appearance of the remains of their ancestors in earth formations, the study of which constitutes Fossil Botany or Paleo-Botany. A knowledge of the geographical distribution of plants is of great service in the same direction, and this constitutes Geographical Botany. As the genesis of any particular kind of plant is frequently indicated in the early states through which an individual of that kind passes, the truths of Vegetable Embryology are of great assistance to the systematist. Each of these divisions may in turn become the principal object of pursuit and all the others become tributary to it. Each also has its peculiar phases and its subdivisions. For example, when structural botany has for its object merely the determination of the organs as they exist it becomes Organography. When such object is to determine the development of organs through the transformations of others it becomes, in a restricted sense, Organogeny or Morphology.

The ancestral organ and its developed

product are called Homologúes of each other, and a Homology or Affinity is said to exist between them. When they are only similar, without any genetic relationship, they are Analogues of each other, and Analogy exists between them. Morphology might, therefore, be defined as the study of homologies and analogies. As classification has been based very largely upon flower structure and fruit structure the study of these, respectively, has been dignified by the titles Anthology and Carpology.*

The description of plants in such manner that they can be recognized therefrom is called Descriptive Botany or Phytography. Botany has also numerous departments depending upon the class of plants under study, as Phanerogamic Botany, the botany of flowering plants: Cryptogamic Botany, that of flowerless plants; Mycology, the study of fungi; Agrostology, the study of grasses.

DEPARTMENTS OF ECONOMIC BOTANY.

The mere fact that botanical knowledge is utilized in a certain art is not sufficient to establish a separate economic department of the subject. This depends upon its application in such a way as to demand and receive special researches of a scientific nature in the performance. The history of the arts shows that very rarely have the investigations of scientific botany proceeded far enough or its deductions been so presented, that the economic botanist has been enabled to depend upon them alone. Almost invariably has he been obliged to extend and modify them in special directions. Upon such facts depends our recognition of the absolute necessity of scientific study and training as a foundation for high art in any department. The departments of economic

*The classification of plants upon the basis of anthology and carpology alone has prevailed largely in the past. The more recent development of the department of minute anatomy and even of physiology has indicated an importance not previously suspected of the facts so ascertained in establishing relationships, and systematic work which does not take due cognizance of these subjects is not likely to be accepted in future. At the same time it is to be noted that the results so reached are with but few exceptions only corroborative of the deductions of the departments above mentioned.

botany are named from the particular arts for the advancement of which they are established. Agricultural Botany is subdivided into a number of different departments, such as Agrostology, or Graminology, the study of grasses and of their culture; Horticulture, the study of garden plants and of their culture; Floriculture, Pomology and Forestry. Doubtless a very large and important department will yet be established in the study of medicinal plants and of their culture. The term Medical Botany is self-explanatory as to its general nature. In use, however, it should be more strictly regarded than is customary. The term originally included all botany relating to medicinal plants; but with the development of Pharmacy the greater portion of what was once comprised in the former term has naturally and permanently established itself in the form of the separate department Pharmaceutical Botany. Medical Botany properly concerns itself with the medicinal properties of plants, the localization of the proximate principles of the latter, their uses, including the principles (but not the practice) of their preparations as based upon such facts, and their classification in view of medical considerations.

PHARMACEUTICAL BOTANY.

In its widest scope Pharmaceutical Botany would include the classification, phytography, histology, distribution and culture of medicinal plants; the collection, preservation, packing, transport, commerce, identification and selection, composition and methods and processes of preparation for use of the drugs derived from them. From this it would follow that the pursuit of pharmaceutical botany would demand a thorough knowledge of nearly all departments of scientific botany. This conclusion is to be modified, in view of existing conditions, in important directions. The pursuit of the study to such an extent would almost necessarily involve the average pharmacist, at least in this country, in financial failure, through the inattention to practical affairs which would ensue. It is the peculiar office of the teacher of technical science to place its practical benefits within the reach of his students while relieving them from attention to the greater portion of the field. It is not to be overlooked, however, that while such a process of extensive exclusion is possible, utility requires that a corresponding degree of elaboration shall be attained in special directions. The faithful teacher, moreover, will not refrain from urging as liberal an indulgence in extra-utilitarian study as individual circumstances will properly permit. The directions in which botanical knowledge is most useful to practising pharmacists will determine the most important requirements for botanical study. The identification and selection of drugs—that is to say, Pharmacognosy—constitute the principal field for the exercise of botanical knowledge on the part of the pharmacist, and it is those departments of botany which bear directly upon this subject, and upon Materia Medica, to which attention will be restricted in this essay.

It is convenient to divide botanical pharmacognosy, like vegetable anatomy, into the gross and minute, the latter concerning itself with those characters which require the compound microscope for their demonstration. Remembering that vegetable drugs may consist of the entire plant or of any one or more parts thereof, and that they may reach the pharmacist in any condition, from that of unbroken or even fresh to that of a fine powder, the department of botany necessarily pertaining to pharmacognosy and materia medica will appear as follows:—A knowledge of classification or systematic botany, while a prime necessity in medical botany, there being a distinct co-relation between natural classification and medicinal value, is one of the less practical and essential elements of pharmaceutical botany. Still, it aids the student in the application of phytography and especially in understanding distribution, and it serves to crystallize and systematize his knowledge of groups of medicinal agents. A good working knowledge of phytography may be regarded as the leading essential. If the drug is to be sought by the pharmacist in nature, he can recognize it only through phytography, whether that knowledge be acquired through folk lore or book lore.

If, on the other hand, he seeks the crude drug in commerce, he merely restricts his phytography to the plant part under inspection, and so far from being by this consideration relieved from phytographical labor, its requirements are the more exacting and its methods the more refined, as the recognition and estimation of a fragmentary representative becomes more difficult than that of the complete individual. As "Phytography" in its ordinary employment is about equivalent to "the study of the manifest organs of plants," or of their gross units of structure, morphology becomes the key to the situation. He who endeavors to understand and familiarize himself with plant structure by regarding the organs in a disconnected and arbitrary view encounters a tedious and uninviting task, and it may truthfully be said that it is this method or custom which has rendered the study of botany very much of a *bete noire* in pharmacy. Proceeding upon the basis of morphology and recognizing in the plant a structural unit through the transformations of which the higher or more recently developed organs have arisen, the study becomes in reality as in name, natural history, and as full of meaning and as entertaining a subject of thought as history of any other class. Such pleasures, however, do not constitute its justification in pharmacy, but rather the fact that by so studying the pharmacist effects a great economy of time and fixes his knowledge beyond the possibility of ready dislodgement. When drugs come to hand in a comminuted condition, as they really do in a vast majority of cases, the compound microscope is the only resource, and the department of plant histology becomes the foundation of work. As will be shown farther on, the greater portion of this subject can be passed over, but that portion which receives attention, permitting the recognition of detached tissue elements and the determination by their examination of their source, requires observations quite as careful and knowledge quite as accurate as are called for in any other portion of the field. In pursuing these observations micro-measurement and micro-chemical manipulation are indispensable, and a lack of ready facility in drawing will be appreciated with regret.

An intelligent knowledge of the distribution of medicinal plants and of the collection, preservation, packing, transport and commerce of drugs manifestly constitutes the highway to a commercial success, and it is to a considerable extent the demands of this portion of the pharmacist's work which, under existing conditions, prevent a closer attention to scientific details.

The cultivation of medicinal plants is as yet in its infancy. Its development will call for a very extensive additional equipment on the part of the pharmaceutical botanist in the direction of plant physiology, this depending in turn upon a great extension of the department of histology in connection with physiological chemistry.

The composition of drugs and the methods and processes of preparing them must be regarded from a chemical standpoint, and it is only by virtue of association that we regard this portion of pharmaceutical chemistry as pertaining to pharmaceutical botany.

Finally we note that only an insignificant portion of the materia medica originates in flowerless plants, so that the great department of cryptogamic botany may with profit be dismissed from our consideration as relates to detailed treatment.

ORDER OF SUBJECTS.

In attempting a comparative view of the series of plants it is unquestionably well to begin with the lowest form and follow the line, or rather lines, of upward development. But in gaining our first knowledge of the structure of the plant organism, sound and accepted rules of pedagogy require that we begin with the more obvious characters of the higher plants, and pursue the analytic method, so far as the special conditions of the case will permit.

It has been repeatedly remarked that plant life is a circle of germination, growth and reproduction, passing again into germination. It therefore makes little difference, on general principles, at which point we enter upon our series of observations. Begin where we will, we must labor at the disadvantage of requiring more or less knowledge of facts preceding our point of departure, and therefore not as yet studied. In special cases, however, there is much room for choice, and there are many reasons why we would advise pharmaceutical students to commence by observing the organ concerned in reproduction, namely, the flower.

Not until mankind shall learn to breathe by some different method from that followed by his ancestors will any method of studying botany become popular which passes over the rose, buttercup and aster for the making of microscopical preparations of myxomycetes, or the sectioning and staining of tissues.

ANTHOLOGY.

THE GENERAL NATURE OF THE FLOWER.

In order to accurately understand the structure of the flower we must first consider the general characters of its structural units, pertaining in turn to the stem upon which the flower is borne and of which it is a part. These are well displayed in a willow twig (Fig. 1), presenting a main stem, with perhaps short branches below and leaves above. These leaves are found upon examination to arise at regularly occurring points, thus dividing the stem into parts which are seen to possess definite and uniform characteristics, each being complete in itself. In common language, these parts are called "joints" and

technically, Phytomers. The upper portion of each is commonly somewhat enlarged and it possesses the power of giving rise to three new structures:—(1) the leaf (a), or in many plants a circle of two or more leaves; (2) a superimposed phytomer, continuing the growth of the stem in the direction of its original axis; (3) a branch (b), extending the growth of the stem in a lateral direction, or, if there be more than one leaf, then a corresponding number of such branches. Upon the upper portion of the stem the branches are seen still undeveloped, and in the form of buds (b). The bud originates, with rare

Fig. 1.

2. There is a definite and regular arrangement as to position of the leaves upon the stem in most cases.

3. Several leaves and as many branches may develop from one node.

4. The branch normally develops in the leaf-axil, and conversely a leaf, in some form, is normally at the base of each branch in its rudimentary condition.

5. All growth developing in the leaf-axil is a manifestation of the branch.

6. All organs of the plant which we consider, except the root, are constructed of the above parts in some modified form, or are appendages upon them.

Certain necessary qualifications of these statements can be made only when we come to the study of the stem, and these do not involve any failure to understand correctly the principles of anthology.

Before proceeding to consider the forms of structural modification of the phytomers in the development of the flower certain important properties pertaining to them, in addition to their ability to multiply and grow as above indicated, should receive attention in order that later a comparison of reproductive methods can be instituted. It is found that if, in the case of certain plants, a stem be laid prostrate in the soil, its connection with the parent not destroyed (Fig. 2.), its nodes, in addition to producing branches (a), may develop roots (b) similar in structure and

exceptions, at the point where the leaf emerges from the stem and upon the upper side of the former. This point is known as the leaf-axil. The portion of the phytomer which gives origin to these three structures is called its node (c). The portion intervening between two nodes is called the internode (d). The internode does not normally possess the power of giving origin to new parts. The branch is found, after development, not to differ essentially from the stem, so that a branch may be regarded as a lateral stem, secondary, tertiary, and so on. In noting hereafter the development of the other parts of the plant out of those here named we shall frequently find the latter so modified that we shall be unable to recognize them by the ordinary methods of examination, and the relative positions which they occupy will prove an important guide. A correct understanding of morphology requires, therefore, that we keep in mind the following facts relating to the internode, node, leaf, branch and superimposed phytomer.

1. Any of them may remain more or less undeveloped.

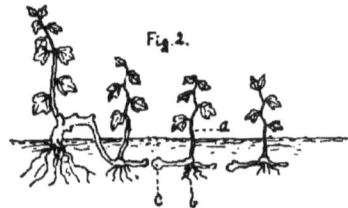

Fig. 2.

function to those of the parent. If now the phytomers be separated through some portion of the internode, they will heal the wound so produced by the formation of a callus (c), continue to grow independently and become plants similar to the parent. Such a process, here of artificial production, is of frequent natural occurrence and is called Propagation. It is seen to be purely vegetative, and may be defined as the production by vegetative processes of a plant body growing independently and separately from that from which it was derived. The question has been raised as to when the new plant becomes an individual, or whether it does so at all. Their conduct after separation shows clearly that they previously possessed perfectly independent powers of support. If an individual after separation, why not before,

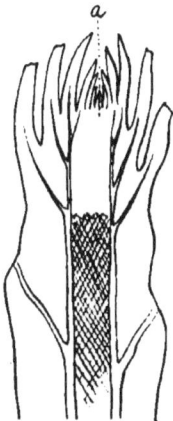

Fig. 4.   Fig. 3.

and why is not each phytomer upon the plant an individual? If, upon the other hand, it be considered that even separation does not involve individuality, then the vast number of propagated plants which may result from an original seed must constitute but a single individual.

Various other modes of stem propagation may here be referred to, and it may be remarked that the process is not confined to the node, occurring in exceptional cases from fragments of the internode, root, or even leaves. The phytomers, instead of remaining attached during the rooting process (Layering) may be first separated (Propagating by Cuttings). The cutting, then called a Scion, may be inserted (grafting) or a bud may be so inserted (Budding) under the bark of a living stem, or may be caused to take root in the soil. Propagation by tubers or parts of them, as in the case of the potato, is identical. It may be remarked in passing, that in the seed itself nature resorts to a similar method, for the contained embryo consists of one or more phytomers.

Roughly stated, the phytomer may be said to consist of three portions: (1) a framework consisting of strands of conducting vessels and commonly of fibres; (2) between and around the last a quantity of soft, non-fibrous tissue; (3) a covering, membranaceous when young and changing greatly with age. All these parts are extended into the leaf, the first existing in a system of branching ribs or veins, the second as a filling in the meshes of the former, and the third as a highly developed epidermis. Morphologically considered, the typical leaf (Fig. 3) consists of three parts which, like those of the stem, will be considered in detail hereafter. The Base (a) bears the Pulvinus or organ of attachment to the stem, and upon either side a membranous expansion (b) called the Stipule. The stem of the leaf (c) is called the Petiole. The blade (d) is called the Lamina. In some plants an additional organ, the Ligule, develops as an appendage upon the face.

If we could observe the phytomers of such a twig during the process of formation in the bud (Fig. 4) we should find them in a more and more rudimentary condition toward its apex or centre until we reached an ultimate growing point (a), where development had not yet manifested itself. Yet this point would possess the power, under proper conditions, of continuing indefinitely the process of development and growth of phytomers. It therefore may be said to represent a certain amount of vital energy or potential growth, though by this it is not meant that such energy is actually stored there. Now, our fundamental ideas of flower structure rest upon the fact that this vital energy or potential growth may be diverted from the production of phytomers and leaves such as we have been considering and may produce in their stead other structures in which resemblance to and variation from them are mingled in variable proportions. These new structures we then call Modified Phytomers and Modified Leaves. The student should here dwell upon this point until the exact meaning of these terms becomes clear. When hereafter he encounters, as he very frequently will, a reference to some organ being modified or transformed, it must never be understood that it was first produced and then changed. The exact meaning is that the change takes place in the direction or exercise of the energy which is to produce the modified structure.

Such a diversion of energy may be caused by accident, as seen in the so-called "Willow-cone" (Fig. 5), resulting from an injury inflicted by an insect in depositing its eggs in the centre of a bud. A portion of the structures, having been originated before such injury, will reach a partial development, but further production is checked and a distorted product results.

In the case which we have to consider the modification dates from an earlier stage and is natural and physiological, instead of pathological, as in the case of the willow-cone. Fig. 6 represents one twig after the fall of its leaves in the autumn. Each bud is seen protected by its lowest leaf, permanently enlarged and developed into a covering scale. At the base is seen the scar of the leaf in the axil of which the bud was developed. Fig. 7 illustrates the twig in the spring after early growth has enlarged the buds. In Fig. 8a the covering scale has fallen, the branch has developed, and its structure can be seen to consist of a great number of very short phytomers, each of the crowded nodes bearing a scale (c) and in its axis (Fig. 9) a peculiarly shaped body. That the scales are modified leaves is proven not only by their position, as previously explained, and further explained in our study of the leaf, but by the fact that in exceptional cases the branch will produce them in a form intermediate between that of a scale and of an ordinary leaf (Fig. 10, a). Such being the case, anything produced in their axils must, according to the same laws of position, be modified branches. We must therefore regard the body shown in Fig 9 as a modified branch, one of a

great many produced upon the parent modified branch shown in Fig. 8. How profound is the modification which has taken place in the latter can be appreciated from a consideration of its reduced size, for it is now approximately full grown. The great number of phytomers upon it, had they reached the form and extent of development reached by those in Fig. 1, would have produced a branch many feet, or even yards, in length, whereas in their present form they will produce a structure only an inch or two long. As we shall soon see, increased complexity of structure has replaced the greater amount of tissue growth of the leafy branch.

Fig. 5.   Fig. 6.   Fig. 7.

Examining now the little modified branch (Fig. 9) we observe that it presents two uniform portions or halves, united into a single body except at the tip, where they are separate. In exceptional cases we find this separation extended downward, perhaps even to the base of the body, and each of the separated portions expanded, formed and veined very much like a small leaf, which, in fact, it is. The little branch a-d is then to be regarded as bearing two leaves which have been developed in a united condition. Upon dissection (Fig. 11) the body thus

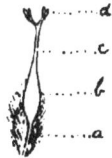

Fig. 8

Fig. 9

Fig. 10.

formed is found to be hollow at one portion, bearing two slight projections upon its inner wall (a), and upon these a number of minute rounded bodies (b). If allowed to develop and mature under the requisite conditions we should find that these bodies had become seeds. The structure producing them we now see to be *a reduced branch, modified for the production of seeds, and this constitutes our definition of The Flower*

It does not follow that because constructed for the production of seeds a flower is always capable of performing this office independently, and, indeed, such is not the case with the flower under consideration, which is an Imperfect one. Its minute structure, to be described in a succeeding chapter, shows it to contain within the bodies which are to become seeds, minute structures called Macrospores, which produce cells comparable, in their essential characters, to the ova of animals, and requiring a similar fertilizing process to cause their development. Flowers, or at least certain of their products, are thus seen to possess sex and to be capable of performing sexual reproduction, or Reproduction Proper. Commonly both sexual parts are present in one flower, and of these the female is called the Gynaecium, frequently represented by the symbol G. The gynaecium, or one of its parts when these are entirely separate from each other, is called a Pistil. This flower possesses only

gynaecium, and is therefore often spoken of as a "Female Flower," technically Pistillate, and indicated by the symbol ♀ This gynaecium consists of but a sin-

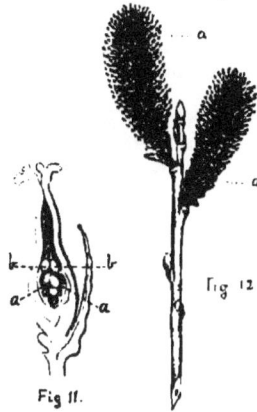

Fig 12.

Fig 11.

Fig. 13

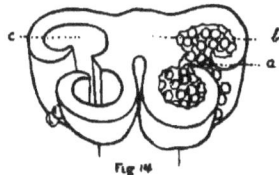

Fig 14

gle pistil, but the number of pistils a gynaecium may possess is indefinite.

Before considering the structure of the pistil we will examine a "Male Flower," borne, in the case of the willow, upon a plant which produces no pistillate flowers. Fig. 12 illustrates branches (a) crowded with male flowers, each (Fig. 13, a) in the axil of a scale (b). In this case the two modified leaves forming the flower are entirely separate and the hollow portion of each (d) is small, borne at the summit of a stem (c) and filled (Fig. 14) with a great number of minute rounded bodies. These correspond, though of the other sex, to the macrospores which we have found the pistillate flower to produce, and they are called Microspores, in flowering plants called Pollen-grains. They possess the power of germinating, growing and producing Male Cells, comparable to the spermatozoids of animals, and requisite for the fertilization of the corresponding element produced by the macrospores. The male portion of a flower is called the Androecium, frequently indicated by the symbol A, and it consists of one or more Stamens, in this case of two. As this flower consists only of androecium it is known as a Staminate Flower, indicated by the symbol ♂

A vessel in which one or more spores is developed is called a Sporangium, Macrosporangium, or Microsporangium, as the case may be, and a modified leaf producing sporangia is a Sporophyll. A plant body bearing spores is called a Sporophyte. When, as in this case, the macrospores are produced by one plant and the microspores by another, the plant is Dioecious. If in addition each plant produced some perfect flowers it would be Dioeciously Polygamous. If, as in the Alder (Fig. 15) pistillate flowers (a) and staminate flowers (b), or, otherwise stated, spores of both sexes, are produced by the same plant, it is Monoecious. If in addition the plant bear some perfect flowers it is Monoeciously Polygamous. When, as illustrated in Figs. 22, 24, &c., the flower possesses both gynaecium and androecium it is Hermaphrodite, indicated by the symbol ☿ Hermaphrodite flowers are not always perfect, as one of the organs, while perfect in form, may be functionless. Imperfect flowers present all intermediate grades between that last mentioned and that in which there re-

mains no trace of the lost part, or in which it has even been transformed into an organ of a different kind.

The parts of the stamen and pistil will now be named.

Of the Stamen —The stem-like portion (c) regarded as corresponding to the petiole of the sporophyll, is the Filament. The portion containing the spores or pollen is the Anther (d). The two halves of the anther, each corresponding to a half of the lamina of the sporophyll, are the Thecae (sing., Theca) (Fig. 14, a). At an

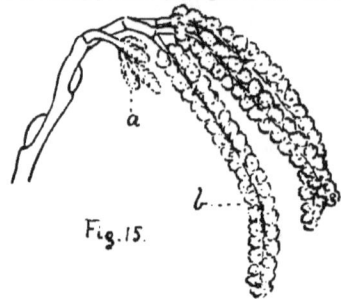

Fig. 15.

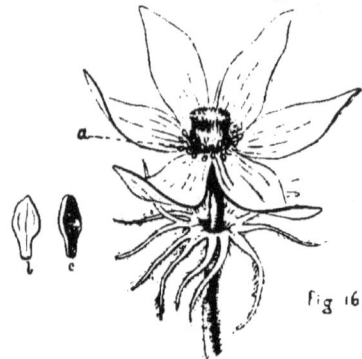

Fig 16.

earlier stage each theca is subdivided into two Locelli (Fig. 14c), and in many plants this condition persists to maturity. The portion connecting the thecae with one another and with the filament is the Connective (b). Our detailed study of the stamen, as of the pistil, may here be anticipated by the statement that any or all of their parts may in other flowers be found modified in an extreme degree by reduction, exaggeration or special form of growth, and may bear appendages in great variety, their true nature in many cases being thus masked. In many cases an appendage apparently consisting of a modified stipule exists.

Of the Pistil —The stem-like base (Fig. 9 a), not present in most flowers, is the Stipe or Thecaphore. It represents the united petioles of the sporophylls. The body of the pistil represents either a single sporophyll having its edges brought together and united with the upper leaf-surface inside of the cavity, or, as in this case, more than one sporophyll with the edges of one meeting those of the other in the same manner, or in many cases in

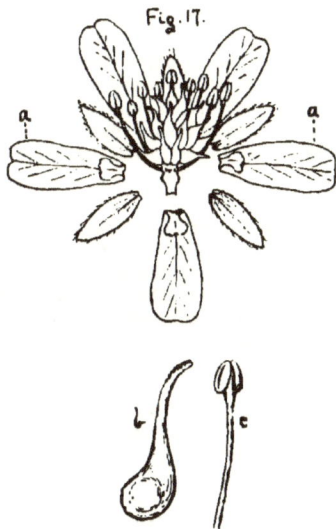

Fig. 17.

a different manner. The edges along the hollow portion, after meeting, project inward more or less. Elsewhere, for a greater or less distance, they may be everted. A sporophyll of a pistil is a Carpel.

The seed rudiments which produce and contain the macrospores are the Ovules (Fig. 11 b). The outgrowth from the inner wall of the ovary upon which the ovules develop is the Placenta (a). The hollow portion of the pistil, containing the placentae and ovules is the Ovary (Fig. 9 b). The divisions of the ovarian cavity, which sometimes exist, are called Cells (Fig. 198, &c.), and the partitions which separate them are called Septa or Cell-Walls. A point upon a pistil (Fig. 9 d), which lacks its epidermis and permits entrance into the ovary of the pollen-product is a Stigma. A usually non-hollow portion connecting the stigma to the ovary and narrower than the latter is the Style (Fig. 9 c).

Since the androecium and gynaecium are capable of producing seeds without the necessity for other floral parts they are commonly known as the Essential Organs, other parts as the Non-Essential Organs.

The danger of accident as the result of blows, punctures, erosion, or even changes of temperature, to the complex mechanism and delicate structure of the essential organs, and the resulting necessity for their protection, is obvious. In the case under consideration the flowers are so closely crowded upon their supporting branch that their subtending leaf-scales afford the necessary protection. But commonly this is not the case, and the flower must provide its own protecting organs, if it have any. It must be borne in mind, however, that protection is the least important office which such organs fulfil.

A series, or apparent or real circle, of such modified leaves, underneath or surrounding the androecium, is displayed in the flower of Pulsatilla (Fig. 16) and constitutes its Calyx, frequently indicated by the symbol K, the several leaves being called Sepals, or Calyx-Lobes, in accordance with conditions to be considered hereafter. Commonly there is a second circle between the calyx and androecium, as in the buttercup (Fig. 17 a), and this is called the Corolla, frequently indicated by the symbol C, its several leaves Petals or Corolla-Lobes, according to their condition. Rare cases occur in which, although but a single circle is present, it is regarded as a corolla. The space between two adjacent petals or corolla-lobes—and the same is true of a similar space between any two organs or divisions standing side by side—is called the Sinus. Occasionally the petals will be numerous, forming more than one circle. Petals and sepals are normally not composed of distinct parts, unless it be by a narrowed insertion, called the Unguis or Claw, which is frequently present (Fig. 18 a), the broad part being called the Lamina, Blade or Limb. They are then said to be Unguiculate. Usually their form is more

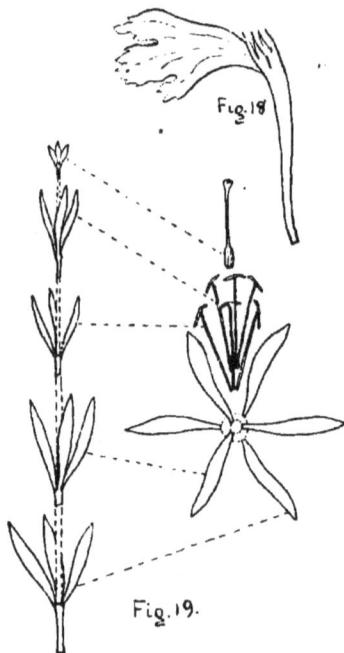

Fig. 18

Fig. 19.

obviously foliaceous than that of the stamens and carpels, and frequently in color and texture, particularly of the sepals, they are also foliaceous. The calyx and corolla may, however, possess

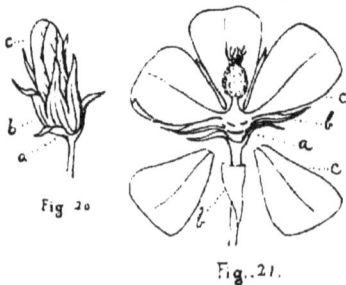

calyx and corolla together when both exist, constitute the Perigone, less aptly called the Perianth or Floral Envelopes. A flower possessing both calyx and corolla is called Dichlamydeous; one with calyx only, Monochlamydeous, indicated by Co, and one with neither, Achlamydeous or Naked, indicated by Ko-Co. Those which are not dichlamydeous are called Apetalous. A flower possessing calyx, corolla, androecium and gynaecium is called Complete. Some plants habitually produce a portion of their flowers without essential organs (Fig 23 a). Such flowers are called Neutral. It must ever be borne in mind that all these parts are constructed of the modified leaves of the floral branch. The latter is called the Torus or Thalamus, or, less desirably, the Receptacle. The torus may, therefore, be defined as *the reduced branch which gives origin to the parts of the flower* Figs. 16 a, 22 and 23.)

The relation of these parts to their branch may be displayed by comparing the leafy stem of a lily with the dissection of a lily flower (Fig. 19). What appears to be a double calyx, or one calyx outside of another, is frequently seen. This appearance is sometimes due to the actual manifestation of two circles, as in the mustard; at others to appendaging (see Fig. 31); sometimes to a circle of modified foliage leaves standing close to the torus (Figs. 20 and 21 a), and known as the Epicalyx. When, as in the last case, the flower has in addition a calyx and

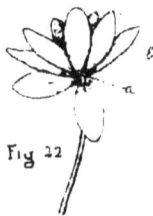

Fig 20

Fig. 21.

Fig 22

Fig. 23.

any color or texture and they may be similar or dissimilar, usually the latter, in this feature. The color and texture of the petals, as of the sepals, may even differ among themselves. The calyx, or the

corolla the real nature of the epicalyx is readily understood. But when (Fig. 22) there is no corolla, the calyx (b) being colored like one, the epicalyx (c) may easily be mistaken for a calyx. In this instance,

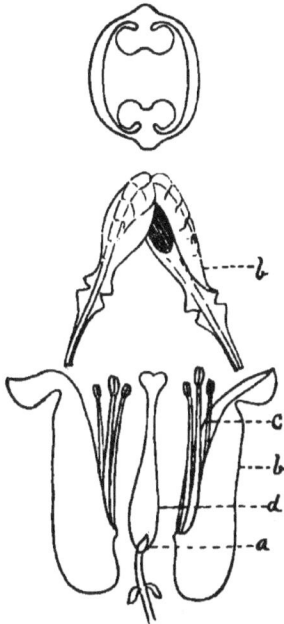

Fig. 24.

however, it may be seen by turning back the epicalyx or removing the calyx (Fig. 23) that its point of insertion is upon the stem below the torus (a), so that it can be no part of the flower proper. The divisions of the epicalyx are called Bracts, though the term is not restricted to this use, as will be seen further on.

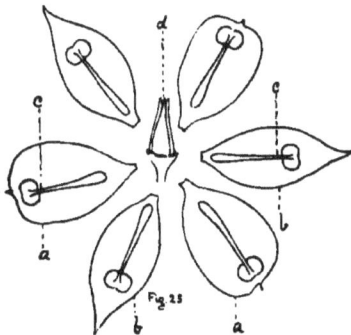

Fig. 25

THE MODIFICATIONS OF FLORAL STRUCTURE CLASSIFIED.

The typical flower, reference here being restricted to its highest forms, will obey certain well defined laws of structure, as regards the following characters: The relative number and position of the parts of different kinds or of different series; the separation of each part from all others of the same or of a different kind or series; a similarity in form and size of the parts composing any one series; the characteristic forms and functions of the parts of different kinds. For the identification of the parts of such a flower the illustrations and definitions already given will prove ample for the beginner. But, unfortunately, such flowers are very rare. The great majority of them deviate from the type in one or more directions to such a degree and in such variety as to very frequently create difficulty in identifying or circumscribing the several parts. To fit the student for properly meeting the difficulties which so arise, as well as

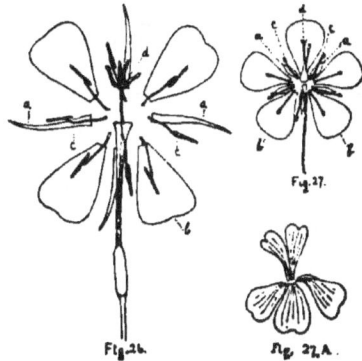

Fig. 27.

Fig. 26.   Fig. 27.A.

for understanding botanical terminology, it is necessary to specify and explain the principal forms of variation and to establish such a classification of them as their varied nature will permit.

The several laws governing the typical flower here referred to may be presented as follows:

LAW 1.—*The parts of each kind or series are either of the same number or a multiple of that number* In the case of the gynaecium it is the carpels which are counted as parts, whether developed each

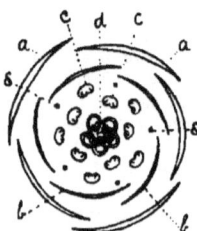
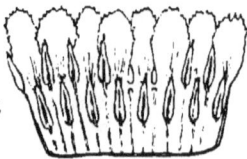

Fig. 28.   Fig. 28a.   Fig. 29.   Fig. 30.

as a separate pistil or all united into one. The number of stamens is normally twice that of each of the other kinds—that is, they form two circles, whatever the explanation of their origin. A flower constructed in accordance with this law is said to be Symmetrical. Thus the flower of Hippurus (Fig. 29 and, in longitudinal section, Fig. 30) has an entire calyx, no corolla, one stamen and one carpel, and is, at least as to its present state, 1-merous or mono-merous. The symmetrical flower of the Bicuculla (Fig. 24) possesses two sepals (a), 4 petals (b), or possibly 4 sepals and 2 petals, 6 stamens (c) and a 2-carpelled pistil (d), and is said to be 2-merous or di-merous. That of the Veratrum (Fig. 25) is similarly based on the plan of 3, and is called 3-merous or tri-merous. Oenothera (Fig. 26) is 4-merous or tetra-merous. The Geranium (Fig. 27) is 5-merous or penta-merous. Fig. 28 displays the plan of the Geranium flower, as seen in cross section, and admirably illustrates the law to be next considered. The term Isomerous is used to indicate that the same number of parts enter into the formation of the two or more circles to which the term is applied.

There are several forms of deviation from this law. Through Suppression there may have come to be too few parts and through Duplication there may have come to be too many. Of each variation there are two forms. In the monochlamydeous flower of Pulsatilla (Fig. 16) suppression of the entire petal-circle has occurred. In the staminate and pistillate flowers of the willow all except a single circle are suppressed. In the Claytonia (Fig. 40) one complete stamen circle is wanting. Such forms constitute Regular Suppression. Irregular Suppression is displayed in the calyx of the Claytonia (Fig. 40), with three of its five sepals wanting, the androecium of the 4 or 5 merous flower of Horse-chestnut, which usually lacks 1 to 3 of the requisite number of stamens, and of the gynaecium of the saxifrage, which has but 2 carpels. In the flower of the common olive both forms appear to have occurred, for only 2 of its 8 stamens remain. To irregular suppression the term Abortion has been applied, while by others this is restricted to suppression in which a vestige of the lost organ remains.

Regular duplication is seen in the 5-merous flower of the strawberry (Fig. 31), with its 10 sepals; the 3-merous flowers of Magnolia (Fig. 32), with 6 to 9 petals, and of Menispermum, with 12 to 24 stamens, and in the 5-merous flower of Malva, which frequently has 10 or more carpels. Irregular Duplication is exhibited in the androecium of the Mustard (Fig. 33), where the multiplication of 2 of the stamens into 4 has occurred. This result has occurred by the dividing up of each into two. The division of an organ into two or more is called Chorisis. We assume that the mass of tissue forming the part should have developed entire, but that at certain points upon its margin separated points have protruded. If growth within each of these points shall not continue, the growth being restricted to the unseparated mass below, then at the maturity of the organ they will still preserve the appearance of more or less small lobes, teeth or protuberances. But if, on the contrary, the growth below

Fig.31.

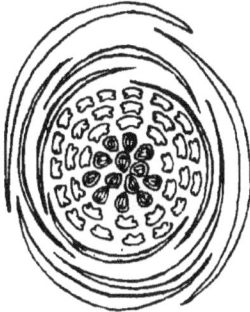

Fig. 32.

Fig. 33.

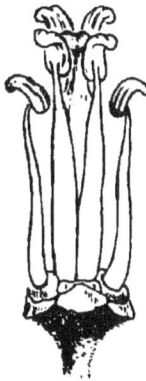

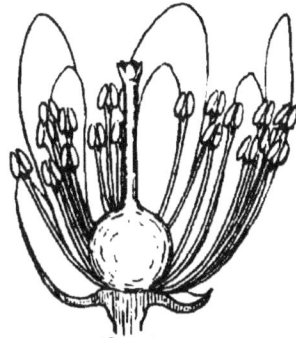

Fig. 34.

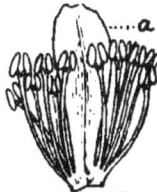

Fig.35.

Fig.36.

Fig.37.

Fig. 38.

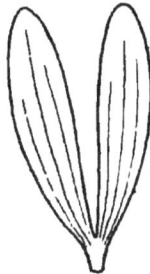

Fig. 39

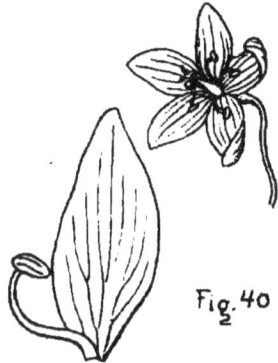

Fig. 40

Fig. 41.

shall cease, and that within the separated parts shall continue, they must become larger and larger, until at maturity the organ shall be seen to consist of a number of more or less completely separate parts. The process is a form of branching, and is usually indicated by the cohesion at the base of the organs thus newly formed, and the location of the bunch in the position normally occupied by one. This is well shown in the flower of Tilia (Fig. 34) and in that of the Hypericum

(Fig. 36) in which vestiges of the lost stamens still remain, as small gland-like lodies (a). Chorisis is well displayed in the calyx of a floret of the Dandelion (Fig. 37), whose sepals have become divided into numerous bristle-like portions (a), and in the corolla of the Stellaria (Fig. 38), each of whose petals (Fig. 39) has become divided into two.

Chorisis sometimes produces an organ

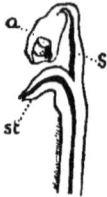

Fig 42.   Fig. 43.

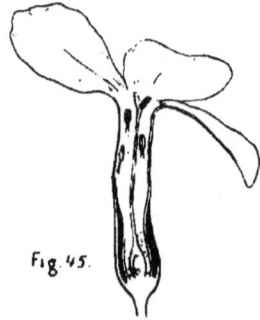

Fig. 45.

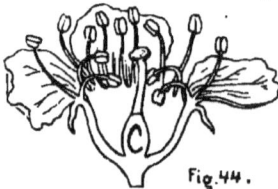

Fig. 44.

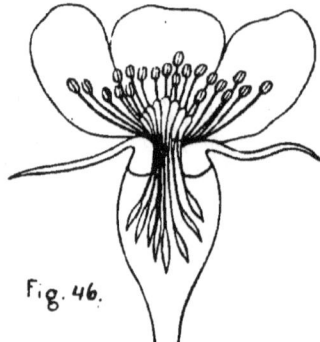

Fig. 46.

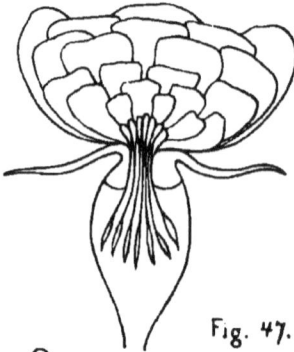

Fig. 47.

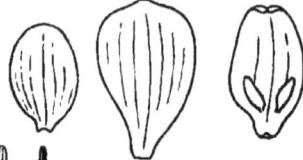

Fig. 48.

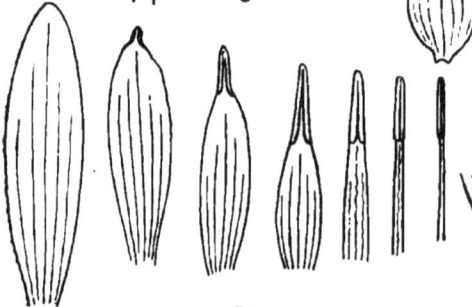

Fig. 49.

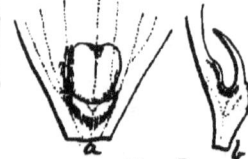

Fig. 50.

of a different kind from the original, as in the case of the stamen of the Tilia, where each of the 3 stamens left after suppression has divided into about 7, and has at the same time yielded one or more little petals (Fig. 35 a) standing in front of the group. Chorisis is thus seen to be Median (in the case of the last mentioned petals) or Lateral (in the case of the stamens), according to whether the line of separation passes tangentially or radially. Chorisis may result in regular duplication (Stellaria) or irregular (Tilia). Regular duplication usually results from metamorphosis, to be considered under law 5.

When the number of organs of one kind, as of petals, as in the rose (Fig. 47) or of stamens (Fig. 46) exceeds twenty, it is commonly spoken of as Indefinite, indicated by the symbol, ∞ although in most cases it falls within certain definite upper and lower limits which are of diagnostic value.

The numerical plan and deviations therefrom are indicated pictorially by diagrams like that shown in Fig. 27, and also by formulae, for an explanation of which the student should consult text-books.

LAW 2 —*The parts of each circle alternate in position with those of the circle next outside and of that next inside.* In other words, each part of the flower stands opposite a sinus of the adjacent outer and inner circles. It is clear that the parts of the two last mentioned circles must stand opposite each other or in the same radial line (Fig. 28), and this will be true of the parts of all alternating circles. This law is also very prettily illustrated by the two stamen-circles in Fig. 28 a. It is also clear that if two circles shall be brought into juxtaposition by the suppression of an intervening circle, their parts will naturally stand opposed and thus appear to invalidate our second law, as in the case of the stamens and petals of Claytonia (Fig. 40). Cases are even known in which such a condition is supplemented by the addition of an inner circle, whose stamens then stand opposite the carpels. Note should also be taken of the fact already pointed out, that the cluster of organs produced by chorisis corresponds in position with the single part from which it was produced.

The treatment of the subject of position here presented is necessarily superficial and incomplete, owing to our failure to have considered already the subject of leaf arrangement. There is a direct correlation between the arrangement of foliage leaves and the parts of the flower. As the arrangement of the former is sometimes by circles or whorls and sometimes by spirals. it follows that some flowers may be arranged on the former plan (Fig. 19), and such we actually find to be the case. There is no one of the floral series but what at times exhibits in its parts (in most cases when they are numerous) a well marked spiral arrangement. In spite of these important facts the above presentation has been made in view of its great practical utility in pharmacognostical examinations.

The student should appreciate the necessity of examining the actual point of insertion in order to determine the position of an organ. as the deviation in position of its upper portion may easily lead to error.

LAW 3.—*The parts composing one circle are all of the same form and equal size* A flower all of whose circles obey this law is Regular. An illustration is found in the flower of Veratrum (Fig. 25). Irregularity may result from abortion (Fig. 127), appendaging (Fig. 52), or mere variation in form (Fig. 91), or size (Fig. 27 a). Sometimes, as in the last case, the variation is so slight that the student will be in doubt as to its existence, while at other times an accidental variation in an individual plant may suggest irregularity where it is not a characteristic. In cases of doubt the relationship of the plant to others whose flowers are regular or irregular may aid to a decision.

LAW 4 —*Each part of a circle develops separate and disconnected from all others in that and in other circles*

As the mass of tissue forming each of the floral parts becomes isolated and projected from the torus its margins and faces are completely separate from those of all adjacent parts. The law assumes that growth shall continue in the isolated portions, by which process they must continue separate. But this form of growth of the parts does not always occur. Very commonly the point of growth changes or becomes restricted to the basal portion, where the development has not yet separated them from one another. The projection of this undivided or unseparated portion from the torus, and its subsequent growth, must clearly result in the development of a portion of the flower consisting of more than one floral part in union. The component parts are usually indicated by more or less of a separation at the apex of the resulting body.

There is no other direction in which deviation from the type represented in Fig. 41 is so frequent and variable as this, nor in which the results are so far-reaching or call for so extensive a classification and terminology. The deviations are of two classes. When a part is united laterally with another part of the same circle the condition is called Connation, Cohesion, Coalescence or Syngenesis. When connation does not exist the parts are said to be Distinct or Eleutherous. Connation will be discussed in our detailed consideration of the several floral parts. Sometimes the parts stand close together, appearing as though coherent or agglutinate, but are neither. They are then said to be Connivent.

In the second form, called Adnation or Adhesion, one circle is united with another. Strictly speaking, the two last terms are not exactly equivalent, as adhesion includes agglutination, while adnation does not. Agglutination is a state which may be mistaken for adhesion, or more especially for cohesion. It is a clinging together of parts by virtue of some interposed substance, rather than true adnation of tissues.

Adhesion may affect any two or more circles of the flower, and it may affect an entire circle or only one or more of its parts.

Thus Fig. 42 illustrates a petal of the Vanilla, adnate to the gynaecium, while the other petals are free. It is plain that when the calyx and the gynaecium are adnate all the intervening circles must be included in the condition, as in the case of the colocynth (Fig. 43). In this case, as in all cases where one or more circles are adnate to the gynaecium, the free or ununited ends of the parts must lose the appearance of emanating from the torus and must appear to emanate from the gynaecium. They are, therefore, said to be Epigynous. At this point the student should not fail to impress himself with an understanding of the fact that in all such cases the epigynous organs really originate at the torus, and that in a cross-section through the adherent portions the microscope will often demonstrate the tissues of such a part adnate to those of the part from which it appears to emanate. In descriptive phraseology the term "Calyx adherent" always means "adherent to the ovary," or epigynous, even though the words "to the ovary" are omitted. Another mode of stating the same condition is to say "Ovary inferior" or "Calyx superior." It frequently happens that the condition is only partial, when the terms "Half inferior" and "Partly inferior" are employed. There are cases where proper application of any of these terms is doubtful and some perplexity is created.

The insertion of a corolla or an androecium which is adherent to a free calyx (as in the cherry, Fig. 44), or of an androecium adherent to a free corolla, as in the Phlox (Fig. 45), is denominated Perigynous. Organs which are not in any way adherent are denominated Free, and because their insertion is manifestly upon the torus, underneath the gynaecium, they are said to be Hypogynous.

With the stamens adnate to the pistil the flower is said to be Gynandrous (Fig. 42). The body thus formed of the united androecium and gynaecium is technically known as the Column. (See also "Stamen-column.").

Two cases arise in which the student may be led to erroneously assume true epigyny. The first is where the disk enlarges upward, surrounding the gynaecium and adhering to it; the other where the end of the branch is hollowed and the gynaecium sunken into and adnate to it, the other organs being thus elevated above and apparently, but not really, upon the ovary (Figs. 46 and 47).

LAW 5.—*Each part preserves its own function and a characteristic form.*

The forms referred to correspond in general with those which have been indicated above. Deviations from this law are caused by Metamorphosis, Enation. resulting in the true appendaging of an organ, the very similar process of exaggeration in the growth of a part, retardation in its growth, or its suppression or

Fig. 51.

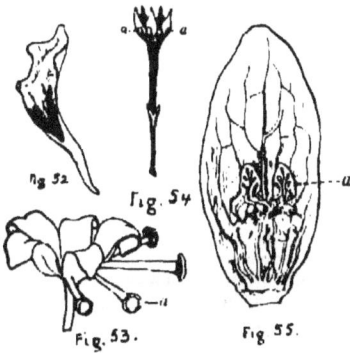

ng 52

Fig. 54

Fig. 53.

Fig 55.

abortion. With the exception of the first the results of these processes will be discussed under the details of the respective parts.

Metamorphosis is the simulation in form or function, or both, of one organ by another. The rose, which normally has but five petals (Fig. 46), is seen under cultivation to consist of a dense mass of them in many circles, becoming a so-called "double" flower. An examination of the inner petals of such a flower (Fig. 47) discloses that they are successively smaller and more stamen-like as they stand nearer the stamens, indicating their origin through the metamorphosis of the latter. which are fewer in proportion as the petals are more numerous. In another form of the rose, the "Green Rose," the petals in turn appear transformed into leaves or leaf-like bodies. Such accidental or artificial deviations from the normal type are called Monstrosities. The

sepals also frequently present a leafy appearance as an abnormality, or in most species habitually. Even the stamens and carpels frequently display the latter abnormality. In all of these cases the change is from a more complex organ, or one of higher rank, to one of a lower, and is called Retrograde Metamorphosis, or Reversion of Type. Progressive Metamorphosis also occurs. It is seen in the gradual transformation of bracts, themselves transformed leaves, into sepals in the Barberry (Fig. 48), and of sepals into petals and petals into stamens (Fig. 49) in the Water Lily. Even stamens may become metamorphosed into carpels and carpels into stamens, one instance being the flowers of the willow, where organs have been seen intermediate in appearance between the two. Cases of abnormal retrograde metamorphosis are very common, and have given rise to a separate department of study known as Teratology.

Enation and the effects produced by it are well illustrated in one of their forms by the petals of certain genera of the Ranunculaceae. The retention of a drop of nectar at the base of the petal of some species of buttercup is effected by the presence there of a minute scale (Fig. 50), covering over a slight depression. The nectar is partly lodged in this pit, partly held between the petal and the scale. In the Coptis (Fig. 51), a closely related plant, the depression is deepened into a more obvious cavity and the scale is dispensed with, while in the Delphinium (Fig. 52) the former becomes a long tube. Although the detailed consideration of

Fig 56.

Fig. 57.

appendages will be taken up in connection with the several organs to which they appertain, we shall here consider a special class of them, called Glands, not only of great importance in diagnosis and classification, but of such physiological importance that they constitute a distinct organ of the flower from that point of view. The peculiarities of structure and secretory function of glandular tissue will be considered in the part relating to histology. Here we note that although glands are sometimes distributed through the other tissues in such a way as to be imperceptible on superficial examination, their tissue is at other times collected into more or less conspicuous bodies of definite form and position. The term "Gland" is frequently applied also to bodies which resemble glands in location and form, but which are not glandular in function. Glands may be stalked (Fig. 53, a), sessile (Fig. 54) or depressed (Fig. 55, see Nectary), and they may develop upon various parts of the flower. Those upon the outside of the calyx are extensively utilized in classification in the natural order Malpighiaceae, while those upon the inside are so used in the order Apocynaceae (Fig. 56) and Gesneriaceae.

Fig. 55 is an illustration of glands located upon the corolla, while Fig. 57 illustrates them connected with the stamens, as seen in Sassafras.

Glands upon filament-like stalks, suitably located, may easily be mistaken for stamens. Glands may themselves be appendaged.

The misleading effects of suppression have been observed in the Hepatica (Fig. 22) in the absence of the corolla, the metamorphosis of the calyx toward corolla and of the epicalyx toward calyx. Those of abortion are seen in the Pulsatilla (Fig. 16), where the petals b and c are reduced to simulate sterile filaments.

Exaggeration of growth is well displayed in the torus of the strawberry and the placentae of the watermelon, which respectively contribute the massive edible portions of those fruits.

The principles of anthology having thus been followed into and through the typical flower, and the general nature of the deviations therefrom having been outlined, we shall proceed to a consideration of the several parts of the flower, with the object of preparing us to interpret the multiform appearances which those organs present in the extensive flora from which our drugs are derived.

That division will not, however, close our consideration of flower structure, as some important modifications will remain to be discussed in our chapter on pollination and fertilization.

THE PERIGONE

The perigone is to be studied as to the number of its circles, their color, texture and surface, the number of parts forming each, their adhesion or cohesion, the form and divisions, if any, of each and of its parts, appendages, secretions, metamorphosis or other variations, arrangement of the parts in the bud, movements or other noteworthy habits and duration. The normal condition of two circles, and its modifications through abortion, suppression, duplication and metamorphosis, and their agreement with the numerical plan of the flower and its modification through the same influences, need no further discussion. The typical idea of a calyx more or less herbaceous and a corolla thin, delicate and brightly colored, is not always realized. In the Crocus and most related flowers the parts of both circles are similarly petaloid. The petals of Garcinia are thick and fleshy, in Caopia they are leathery and in Alzatea hard and almost wood-like, at least when dried. The surfaces of the sepals, particularly the outer, are not commonly glabrous, while those of the petals are; but even the latter are often glandular, pubescent, densely woolly, or even prickly (some species of Solanum). No shade of color is denied to the perigone, nor is the color necessarily uniform among the parts of the circle or even over the surface of any one part. The shade and markings are very liable to vary in different individuals of the same species (varietal or formal variation), so that color is not always a good character on which to base a determination. In general the color deepens as the altitude of the habitat increases. The number of parts entering into either perigone circle, whether these exist in a distinct or a coherent state, is indicated by the appropriate numeral preceding the

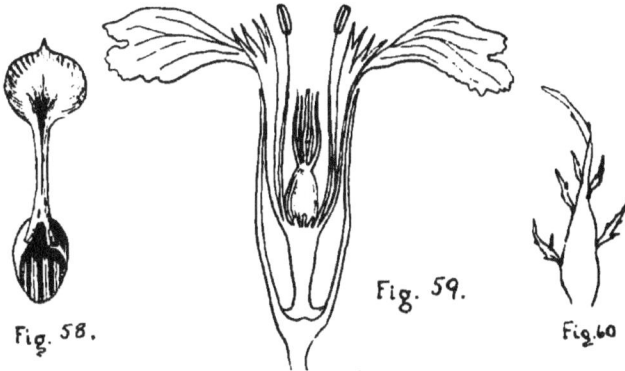

Fig. 58.

Fig. 59.

Fig. 60

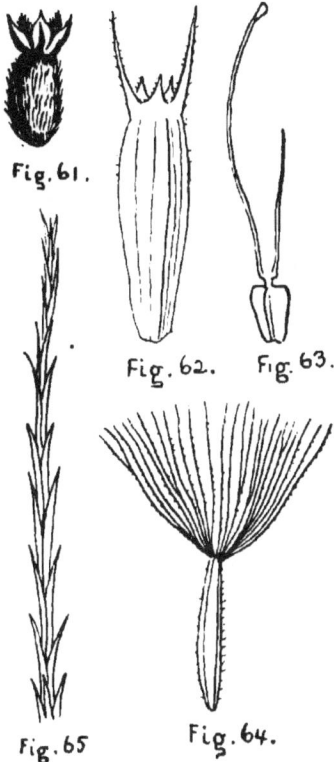

Fig. 61.

Fig. 62.

Fig. 63.

Fig. 65

Fig. 64.

suffix "phyllous"—thus, Monophyllous, Diphyllous, Triphyllous, and so on.

The parts of the calyx commonly present themselves to us in the form of a cir-cle, though sometimes as two (Mustard).

The strictly typical state calls for a general resemblance between the form of the perigone parts and that of the foliage leaves of the plant which bears them. They sometimes display a keel corresponding to the mid-rib of the leaf, and, as in the leaf, this may be continued into a terminal point.

They may be concave, as in Theobroma (Fig 93); the margin may be toothed and the apex toothed or fimbriated, as in Silene (Fig. 18). The toothing of the margin may extend into a pinnatifid condition, as in the calyx lobe of Rosa canina (Fig. 60), and that of a toothed or fimbriate apex into the cleft or divided state of chorisis. The peculiarly divided calyx illustrated in Fig. 64 is denominated a Pappus, and this term has been extended to all forms of the calyx (61 to 68) existing in that family (the Compositae) and in some others. Fig. 68 illustrates the action of median, as well as of lateral, chorisis, in the development of a double pappus. The several forms of perigone parts corresponding to those of leaves (see leaf-forms) and numerous intermediate ones not illustrated, should be carefully considered by the student, as they have a most important bearing upon the forms of the corolla produced by cohesion, which we shall shortly consider.

Both adhesion and cohesion are exceedingly common in the case of the perigone. The former has already been pretty fully considered. Very rarely is it so complete that there is not at least a portion of the parts remaining free. Since the adherent parts are alternating, adhesion necessarily involves the effect of cohesion. In the case of cohesion extended very high, peculiar effects, often puzzling to the beginner, are produced, as in the case of the petals and stamens emanating from the summit of the long calyx-tube in Fig. 26.

Cohesion, like adhesion, may be partial or complete. In its slightest forms, with a mere band of union at the base, it may escape observation, as in the case of the corolla of Styrax (Fig. 71). In such cases a decision is best reached by carefully pulling away the corolla in a lateral direction. If there is union, however slight, the corolla may thus be removed as one body. Upon the other hand, the petals will occasionally be found held together at the base by agglutination, in which case they can be readily separated without the tearing of any tissue. A

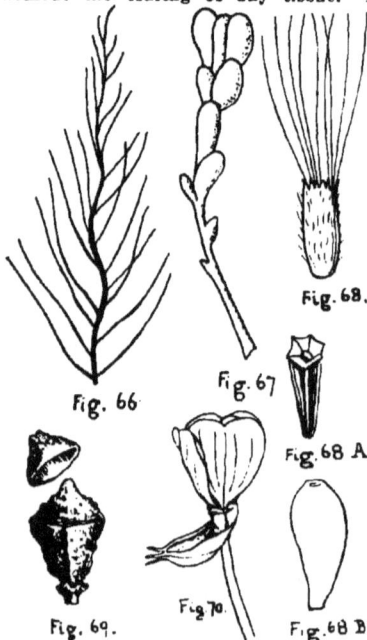

Fig. 68.

Fig. 66

Fig. 67

Fig. 68 A

Fig. 69.

Fig. 70.

Fig. 68 B

peculiar and extreme form of cohesion is that in which the sepals refuse to separate even at the apex when the flower expands, and the calyx is torn loose from

its basal attachment, falling entire as a Calyptra, as in the corolla of Eucalyptus (Fig. 69). A modification of it permits the remainder of the flower to escape

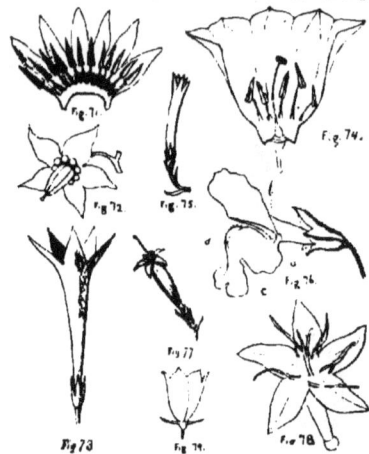

Fig. 71.
Fig. 74.
Fig. 72
Fig. 75.
Fig. 76
Fig. 77
Fig. 73
Fig. 74.
Fig. 78

through a rent in the side, as sometimes in the case of the Oenothera (Fig. 70). When the petals are distinct the corolla is said to be Eleutheropetalous or Choripetalous. The older but less desirable term is Polypetalous. When they are coherent the corolla is said to be Gamopetalous or Synpetalous, the older and less desirable term being Monopetalous. Corresponding terms for the calyx are Eleutherosepalous, Chorisepalous, or Polysepalous and Gamosepalous, Synsepalous or Monosepalous. In the gamopetalous and gamosepalous state the parts cease to be designated petals and sepals and are known respectively as Corolla-Lobes and Calyx-Lobes.

The relative altitude to which the cohesion is carried is indicated by special terms. When existing at the base only the circle is said to be Parted (Fig. 71); when extending about half way up, as in Solanum, Cleft (Fig. 72); when still further, but yet leaving a considerable portion un-united, as in Gelsemium, Lobed (Fig. 73), and when having only traces of the parts un-united, as in Symphytum (Comfrey), Toothed. A peculiar form is that in which the position of the parts is indicated by a mere waving irregularity of the margin, as in the flower of Ipomoea

(Fig. 74), which is then said to be Sinuate or Undulate. The student must not fail to discriminate between the entirely different senses in which these terms are here used, in reference to the entire calyx and corolla, and as used previously in reference to single parts thereof.

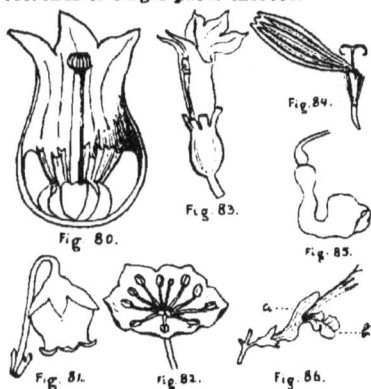

Fig. 80. Fig. 81. Fig. 82. Fig. 83. Fig. 84. Fig. 85. Fig. 86.

We must next consider certain specific forms of the calyx and corolla as wholes, which are of very great diagnostic value. Although such characteristic forms are most numerous among the coherent forms, they are not wanting among those in which cohesion does not exist. Sometimes a non-coherent corolla will necessarily assume such a form through the restraint exercised by coherent sepals. At other times the form is entirely independent of such restraint. Although non-coherent sepals and petals are ordinarily free to assume any position, without regard to their own forms, it is clear that such is not the case when cohesion exists. The coalescence of such petals as that shown in Fig. 18 would necessarily create a long cylindrical tube, without regard to an enveloping calyx, just as has actually taken place in the calyx of that flower (Fig. 59) and in that of Fig. 26. At the same time the union of the broad lobes of either of those corollas would result in a funnel shaped or broader upper portion, as is seen in Fig. 76. In such corollas these two parts are respectively denominated the Tube (a) and the Limb (c). When the change from the narrow to the broad portions of the formative divisions is not abrupt there will be an intermediate por-

tion, as displayed at b in Fig. 76, called the Throat. Less frequently this term is applied also to the delimiting circle between the limb and the tube when these meet abruptly. The terminal boundary-line, including all its extensions and intrusions, is called the Margin. The margin may intrude partly or quite to the tube, so that the cohesion may include none or the whole or any part of the throat, or of the limb. Some of the terms applicable to the forms of the gamopetalous corolla (and, of course, to the gamosepalous calyx) refer to its entire body, while others refer to its several parts. The former class will be first considered. The terms regular and irregular apply to lobes precisely as though they were distinct sepals or petals and to the united portions as well as to the lobes. The regular forms will be first considered. The term Cylindrical is self-explanatory. If nearly cylindrical it is called Cylindraceous. Such shapes are shown in Figs. 26 and 75. If such a one is manifestly angled, as in Mimulus (Fig. 76, calyx), it is Prismatic, and the same

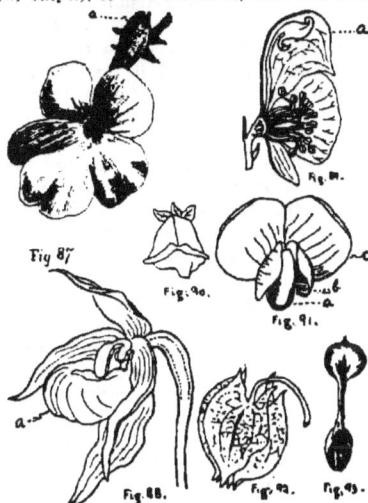

Fig. 87. Fig. 88. Fig. 89. Fig. 90. Fig. 91. Fig. 92. Fig. 93.

is true of other tubular forms. If the entire body flares regularly (Fig. 73), or if there is such a flaring portion upon a cylindrical tube, it is called Infundibular or Funnel-shaped. The less broadened infundibular forms are called Trumpet-shaped, as in the honeysuckle. If the flaring portion or limb is flat, or nearly so, upon a cylindrical or cylindraceous tube, it is called Hypocraterimorphous or Salver-form, as in the flower of the coffee (Fig. 78). A corolla which is bell-shaped is called Campanulate (Fig. 79). Of this there are two sub-forms, the Open and Contracted (Fig. 80). The term Globular or Globose is self-explanatory. It may be specified however, that the mouth must be small and with no conspicuous limb, or with the limb turned back flat against the body. Approaches to the globular form are called Sub-globular or Globoidal. Other related forms are the Ovoid or egg-shaped and Oblong. A somewhat globoidal form, with conspicuous recurved margin, is Urceolate or Urn-shaped (Fig. 81). Of the broader or more widely expanded forms, the campanulate develops outward into the Hemispherical and the Crateriform or Saucer-shaped, as in the Kalmia (Fig. 82). When still more flattened out it becomes Rotate or Wheel-shaped, as in the Solanum (Fig. 72).

A gamopetalous corolla frequently has a fissure on one side extending nearly or entirely to the base, as in the Lobelia (Fig. 83). When in addition the corolla or the split portion of it then loses its tubular form, becoming flattened out, it is called Ligulate or Strap-shaped, as in the Dandelion (Fig. 84).

The applicability to the tube and limb separately of many of the terms here applied to the entire corolla is apparent. It should be noted, however, that very detailed descriptions of these respective parts, as well as of the throat, with specification of any irregularities and marks, are often imperatively demanded. This is especially true in cases of the florets of the Compositae, where such characters, although very slight, frequently serve for specific distinction.

Special terms for forms resulting from the possession of appendages will be considered later.

Terms proceeding from irregularity will next be considered, commencing with those applicable to the entire body.

Either the base or the mouth is Oblique when a plane trans-secting it is not at right angles to the floral axis. The body is declined (Fig. 87) when, either with or without any manifest curve, its axis is turned from the perpendicular, so that it rests more or less against one side of the calyx. It may be Straight or Curved, and the curvature may be Simple or Compound, as in the calyx of Aristolochia (Fig. 85). When dilated upon one side only it is Ventricose, as in some species of Salvia (Fig. 86), or, if the swelling is small and prominent, Gibbous (Fig. 87a). When the swelling is carried downward, so as to form a sac, as in Cypripedium (Fig. 88a) it is called Saccate. When the dilation is directed upward, so as to form a hood, as in Aconite (Fig. 89), it is called Cucullate or Galeate, and when the hood is compressed laterally and much enlarged proportionally to the size of the body it is called Cristate. Most of these terms are also applicable to a single lip of the form next to be considered. When one or more of the lobes of a corolla are separated from the others by a deeper sinus than those of the others it is called Labiate or Lipped. If the fissure proceeds entirely across the corolla, cutting off the lower portion, it becomes One-lipped. Otherwise it is Bilabiate or Two-lipped (Fig. 86). The two lips are denominated respectively the Upper or Inner (a); that which is nearer the stem of the plant when the flower and its stem are standing erect and without any twisting, and the Lower or Outer (b). It is always of importance to note the number of lobes included in each lip, in doing which the student may be misled either by chorisis, one or more extra lobes making their appearance, or by cohesion, two lobes coalescing into one so as to simulate suppression. Two forms of the bilabiate corolla are commonly recognized—the Ringent, in which the lips stand widely apart (Fig. 86), and the Personate, in which the mouth is occluded (Fig. 90).

Several distinctive titles are applied to flower-forms which are characteristic of large and important orders or sub-orders, the Labiate being one. Another is the Papilionaceous, in reference to its simulation of the form of a butterfly (Papilio), as in the common Pea (Fig. 91). The five petals are as follows: Two (a) are more or less coherent by their lower edges to form the Body or Keel; two others (b) are denominated the Wings; the fifth (c) is large, broad and commonly reflexed so as to appear erect, and is called the Vexillum or Standard.

Appendages to the perigone, while less numerous and varied than in the case of the other organs, call for our careful attention, as they sometimes occasion false ·interpretations. In the sense in which the term is here employed we do not refer to hairs and similar outgrowths which modify the surface of the parts, and which pertain equally to other parts of the plant, but to developments which pertain distinctly to the flower, modifying

tended into conspicuous appendages have already been referred to. Sometimes the apex is similarly greatly prolonged into a Cauda or Tail, an extremely exaggerated form of which is sometimes seen. An apex extended into an acute, stiff point is an Arista or Awn (Fig. 63). An awn sometimes emanates from the producing organ at the back, instead of at the apex, and is then called a Dorsal Awn. An awn-like body which is hollow is a Cornu or Horn (Fig. 126, a). A horn-like appendage extending downward is called a Calcar or Spur (Fig. 52). All of the appendages noticed above may be

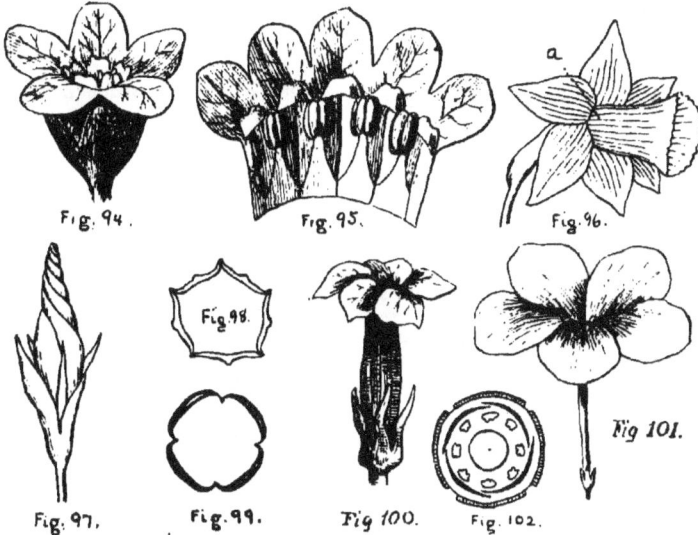

Fig. 94. Fig. 95. Fig. 96.

Fig. 97. Fig. 98. Fig. 99. Fig 100. Fig. 102. Fig 101.

its structure or functions, or commonly both, in some important way. In Nicandra (Fig. 92) we observe a slight appendage at the base of the calyx-lobe on either side and directed downward. Such an appendage, because of its resemblance to the lobe of the ear, is called an Auricle. A similar appendage is sometimes directed upward, and by its union with the contiguous one forms an organ exactly resembling an intermediate or false sepal, as in the Strawberry (Fig. 31). Such appendages, which undergo considerable variation in form and consistency, may or may not be stipular in their nature. Marginal teeth or filaments ex-

found upon either calyx or corolla.

Sacs to the corolla are sometimes intruded, as in Mertensia (Figs. 94 and 95), instead of extruded. The corolla is then said to be Fornicate. Instead of sacs there may be longitudinal folds. When a single large sac occludes the mouth of a bilabiate corolla it is called a Palate (Fig. 76, d). Many appendages such as we have noticed are secretory in function and they may even be glandular in form. Doubtless the various secretions are characteristic, and might, in pharmacy, in exceptional cases be utilized for diagnostic purposes, but the attempt has never yet been made.

Lastly, we note what is perhaps the most important, as it certainly is the most striking and interesting, of the corolla appendages—namely, the Crown. The crown is an outgrowth, more or less membranaceous, from the face of the corolla. Its morphological nature is not understood or agreed upon. It may be a mere abnormal product of median chorisis, or it may be the homologue of the ligule of certain leaves, hereafter to be considered, the latter being regarded as a normal and morphologically distinct part. When it develops from a petal with a distinct narrowed basal portion, which may be assumed to correspond to the petiole of the formative leaf, it usually develops from the point where this is joined to the broader portion, or near it (Fig. 96, a). The crown becomes very important in classification in such families as Passifloraceae and Amaryllidaceae. A ring of intruded folds at the throat is often, perhaps incorrectly, called a crown.

The arrangement of the parts of the perigone in the bud yields some of our most important diagnostic characters as distinguishing orders, sub-orders and genera, and has been the subject of elaborate classification. The demands of pharmacognosy, however, call for attention to only the principal types of Praefloration or Aestivation. The three principal types depend upon the fact that the combined breadth of all the parts of a perigone circle must, (1) be insufficient to enclose the bud, in which case open spaces must be left between their margins (Reseda) or the summit must be left uncovered (Fig. 97), the form in either case being called Open; (2) it must be exactly sufficient to enclose it, the edges then meeting exactly, with nothing to spare, and the form being called Valvate (Fig. 102, the calyx): or (3) it must be excessive, in which the excess may be disposed of in one of several ways. In one, the parts after meeting squarely are uniformly turned straight outward (Fig. 98), the form being called Valvate Reduplicate. In another they are turned straight inward, the Valvate Induplicate form (Fig. 99). They may even be rolled inward, the Involute form; when lapping the one over the other they are Imbricate (Fig. 102, the corolla). Here it is important to note whether the overlapping is from right to left, Dextrorse (Fig. 100), or the reverse, Sinistrorse (Fig. 101).

Other details as to the precise mode of overlapping are frequently worthy of note.

In determining the form of praefloration, care must be taken to select a well formed bud. The praefloration may be mixed, as in Oenothera, where the parts are valvate at the base and slightly imbricate or reduplicate at the immature apex (Fig. 102). At the best, intermediate and perplexing forms will be encountered. Petals, sepals or stamens are occasionally rolled vertically downwards from the apex, this form being called Circinate. Occasionally we find the petals folded and doubled in an irregular manner, the Crumpled or Corrugated form of praefloration. A number of terms are called for by the peculiar conditions of the gamopetalous form. Economy of space is here commonly secured by a longitudinal folding, the Plaited form. Vertical shortening is often secured by twisting, the Convolute form.

The duration of the perianth, especially of the calyx, is frequently of considerable importance from the standpoint of pharmacognosy, although in general not so. When a part falls away at, or very shortly after, expansion, it is Caducous. When lasting about a day, and then either falling or perishing upon the flower, it is Fugacious. When lasting longer than a day, but falling soon after fertilization, it is Deciduous. When remaining and retaining more or less of its normal appearance for some time after fertilization, it is Persistent. When so remaining, but in a withered condition, it is Marcescent. These definitions assume that fertilization takes place normally. If this be artificially prevented or deferred, the freshness of a corolla is often very greatly prolonged. (See Fertilization.) Important facts relating to the Accrescent calyx of the fruit will be presented when the latter is discussed.

Some very interesting facts concerning characteristic movements of the corolla, its sleeping and awaking and other habits, should be sought in general works on botany.

THE ANDROECIUM.

It has already been shown, in considering the general nature of the flower, that the androecium typically consists of two stamen-circles, the stamens of each isomerous with the parts of the other circles, one standing in front of each petal and sepal, that each stamen is entirely free and distinct, and of characteristic form and structure (Figs. 13 and 14). We have also pointed out some of the forms of deviation due to duplication, suppression, adhesion and metamorphosis. To these the following general remarks may be added:—
When an anther, still present, has lost its function, it is called a Sterile or Imperfect Anther. When the anther has become suppressed, but the filament remains, the latter is called a Sterile Filament. One or more complete circles of sterile filaments, changed or not by metamorphosis, may be mistaken for a crown or a disk (Fig. 36). Adhesion of the stamens to the corolla, or even to the ovary, may include only one circle, the other circle being entirely free, or they may be adnate to different degrees (Fig. 28a).

Before discussing other and specific points of variation we shall consider the typical organ in more detail. The number of stamens in the androecium is indicated by joining the appropriate numeral to the suffix "androus," thus, Monandrous, Diandrous, Triandrous, Tetrandrous, Pentandrous, &c. These terms do not necessarily indicate the numerical plan of the flower. When the number is 20 or more, the term Polyandrous or Indefinite is commonly employed.

In color the filament is commonly white or whitish, and the anther yellow, but this is not an absolute rule, as the latter is often blue, brown, black or otherwise colored.

There are several distinct forms of attachment of the anther to its filament which are characteristic of larger or smaller groups. Its origin from the leaf assumes the curving forward and inward of the margins of the blade to become attached to the face of the midrib, producing two thecae, and the production of a secondary or "false" partition separating each theca longitudinally into two locelli. This implies a four-locellate condition of all anthers (Fig. 119). Only rarely is this condition permanent. Ordinarily the false septa more or less completely disappear after the formation of the pollen, leaving the mature anther two-celled, or the condition is brought about in other ways. It is furthermore assumed that the filament is normally continuous along the back of the anther in the relation of the midrib of the formative leaf. This form of attachment is called Adnate (Fig. 103).

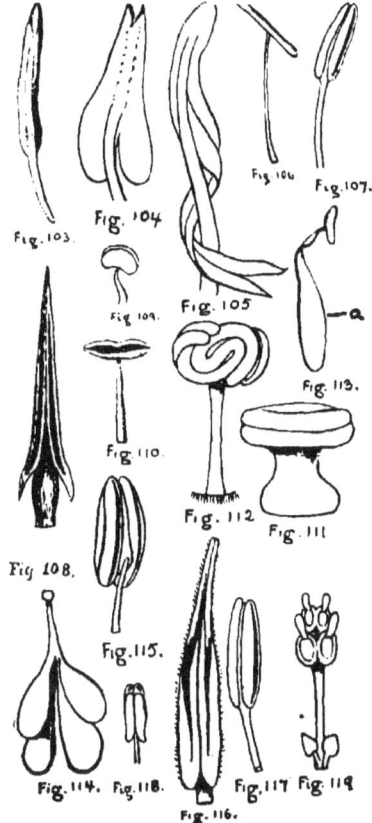

Fig. 103. Fig. 104. Fig. 106. Fig. 107. Fig. 109. Fig. 105. a. Fig. 110. Fig. 113. Fig. 112. Fig. 111. Fig 108. Fig. 115. Fig. 114. Fig. 118. Fig. 117. Fig. 119. Fig. 116.

It may be attached only at some point upon the back (Dorsifixed). Of this there are two forms. In one (Fig. 104) the anther is rigidly fixed, its lower portion close to and parallel with the upper portion of the filament, the Incumbent form. In the other it moves freely upon the pivotal point of attachment (Fig. 106), the Versatile form. Rarely the anther is wrapped or twisted about its filament (Fig. 105).

The continuation of the filament, instead of being along the back, may be centrally up through the base and between the thecae (Fig. 107), the Innate form. The lower portions of the thecae may be separated from one another and from the connective (Fig. 108), the Sagittate form. The sagittate condition is sometimes extreme, the anthers becoming more or less reniform (Fig. 109) or semi-circular, or they may even become horizontal (Fig. 110). This is to be distinguished from the form which is horizontal by versatility (Fig. 106), by the presence in the latter of the two cells side by side, in the former, end to end. Rarely the adnate form will possess the connective upon the inner side (next the pistil), when it is Extrorse by Attachment, in the normal form being Introrse.

Besides these variations in the relation of the two parts, each is in itself subject to certain modifications. Some of them will be discussed in connection with appendages and exaggerated growth. The general form of the filament is subject to much variation which, being characteristic in a given species or genus, requires specification. When cylindrical, either of uniform thickness throughout or regularly tapering, it is Terete. When considerably thickened toward and at the summit, so as to be club-shaped, it is called Clavate. When flattened it is Complanate. Laterally Complanate is so flattened that the edges point toward and from the gynaecium, the broad sides to right and left. Dorsally complanate has the edges pointing to right and left, the broad sides facing toward and from the gynaecium. A dorsally complanate filament may have a sharp ridge or keel running along its back, when it is called Carinate or Keeled. If the ridge is less sharp and prominent it is Costate or Ribbed. It may upon the other hand bear a groove, when it is called Channelled. Rarely a filament is Triangulate in cross-section or otherwise prismatic. When tapering from a broad base to a rather acute apex, it is Subulate or awl-shaped. When very slender or thread-shaped, it is Filiform. When even more slender, hair-like, it is Capillary.

The principal forms of anther are oblong, oval, globular, reniform, quadrangular or linear, and the base or apex may be truncate, rounded, obtuse, acute or pointed. An anther is occasionally doubled upon itself, when it is styled Sinuous (Fig. 112). It may even take the form of a horizontal ring (Fig. 111). This condition is sometimes preceded by the loss of one theca. In any case of curvature, even slight, of the anther, the same is characteristic and of value in classification, as exemplified in the vast genus Solanum, where attention to this character is well nigh indispensable. The filament is also sometimes variously curved or reduplicate, and this condition may be permanent, or only temporary, during the early stage of the flower, for the purpose of utilizing the powerful elasticity of the filament in promoting some function.

Inside of the theca develop certain large cells, the Spore Mother Cells, each of which produces a Tetrad of four pollen-grains. Ordinarily the wall of the mother-cell mostly disappears and leaves the grains separate and mobile, while in other cases they cohere in the tetrad or a cluster of tetrads. Large clusters are called Pollinia or Pollen-Masses. The entire contents of a theca may form one pollinium or they may be divided into several. (Figs. 113 and 114). A cluster of pollinia is called a Pollinarium. The number of pollinia in a theca is of much diagnostic importance in the Orchidaceae. The characteristics of the individual pollen-grains are of the utmost value in pharmacognosy as well as in classification (as, for instance, in the Acanthaceae), and will be discussed in the division of Histology. Pollen-grains represent two classes, with characteristic characters, the one adapted for distribution by insect agency, the other by that of the wind.

We must next consider the structural provisions for permitting the escape of the pollen from the thecae or locellae. This is commonly by splitting, Dehiscence, along a longitudinal line upon each theca, the Suture.

If the suture is at the back of the anther, as in Hyoscyamus (Fig. 115), the dehiscence is called Dorsal. If upon the face, as in the tomato (Fig. 116), Ventral;

if upon the edge, as in Convallaria (Fig. 117), Marginal.

This suture may face the gynaecium, when the anther is Introrse by Dehiscence or away from it, Extrorse by Dehiscence. It does not follow that an anther introrse or extrorse in one way is the same in the other. In the sagittate-horizontal anther

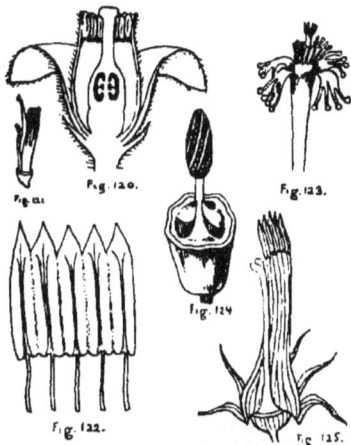

Fig. 120. Fig. 121. Fig. 123. Fig. 124. Fig. 122. Fig. 125.

the sutures of the two thecae often become continuous, the Confluent form (Figs. 109 and 110). Small orifices, called Pores, frequently exist at the apex, as in Cassia (Fig. 118), more rarely at the base. The most scrupulous care must be taken to determine the exact direction in which apical pores look in some cases, as in Solanum, where a slight difference will possess a specific importance. A rare form of discharge is by Valves (Fig. 119), the common form for the four-locellate anther. Special mechanical contrivances for aiding in the discharge of the pollen are of great interest and will be mentioned under Cross-pollination.

Cohesion is responsible for quite as great and important modifications of the androecium as of the corolla. Here as there, it may be complete, or, beginning at either apex or base, it may stop at any point. Fig. 71 displays the dilated bases of the filaments of Styrax lightly coherent, the detection of the condition calling for the same keen inspection as in the case of the corolla. In Guarea (Fig. 120) the union is seen carried to the an-

thers, themselves left distinct. Coherent filaments are styled Monadelphous when all united (Fig. 120), Diadelphous, when there are two groups, even though one of them contains but one stamen, as in Glycyrrhiza (Fig. 121), Triadelphous when three, and so on. It must not be lost sight of that the terms are applied similarly whether the union is progressive, or the result of cohesion, as in this case, or that of incomplete chorisis, as in the Tilia (Fig. 35) and Hypericum (Fig. 36), though its classificatory value is very different in the two cases. The term Column, previously explained, is changed to Stamen-Column for monadelphous stamens. The stamen-column is ordinarily hollow, containing the Gynaecium, but when the flower is staminate it is solid, called a Synandrium. The term Syngenesious is applied in a special sense to coherent anthers, as in Fig. 122. When, as seen in Fig. 72, the anthers come together but do not actually cohere, they are called connivent. The cohesion is carried only partly down the filaments in the squash (Fig. 124), and in the Lobelia (Fig. 125), but in the Asclepias (Fig. 126) it is complete for the entire organs.

A lack of symmetry and regularity, acting separately or together, is responsible for a number of characteristic and important states of the androecium requiring distinctive terms. In the 5-merous flower of Scutellaria (Fig. 127) six stamens are suppressed and the remaining four are irregular, there being a similar pair of each form. This form of androecium has received the title of Didynamous. In this case the anthers of a pair are connivent also. In that of the Mustard (Fig. 33), two of the stamens have each by chorisis become converted into two, these longer than the undivided pair. This form is styled Tetradynamous.

No other subject connected with the androecium calls for such close and discriminating attention in connection with pharmacognosy as the products of exaggerated growth and euation. No portion of the androecium is free from their effects, which apply equally to them when adherent or coherent, free or distinct. The simplest form of appendage to the filament is that of stipuloid appendages to the base, called Petaloid when assuming the form of a petal, as in Fig. 129. A similar appendage may stand in front of a stamen. One standing in front of a sta-

ment at the base may be modified into the most grotesque forms, as shown in Figs. 135 to 137. Such appendages, in any detail of number, form, position and direc-

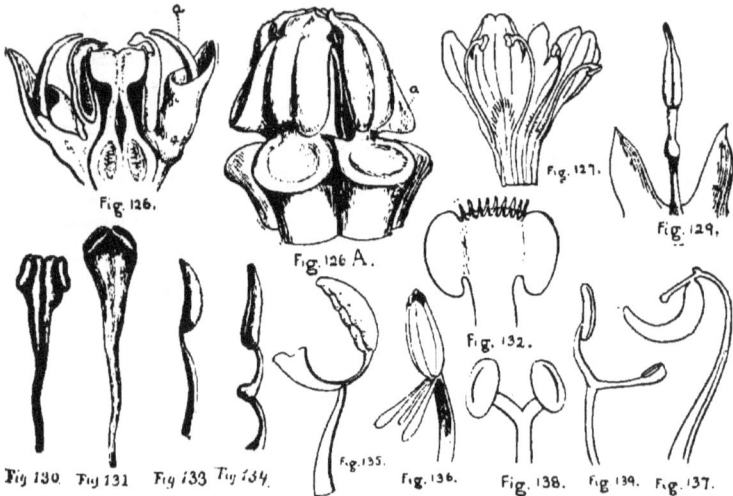

Fig. 126.

Fig. 127.

Fig. 129,

Fig. 126 A.

Fig. 132.

Fig. 130. Fig 131. Fig 133. Fig 134. Fig. 135. Fig. 136. Fig. 138. Fig 139. Fig. 137.

men-group has been shown in Fig. 85. Appendages may be developed at a higher point in other cases. Appendages in the form of teeth or hairs are very common. Modifications of the connective are numerous and remarkable. The thickening of the entire body, equally or unequally, produces such appearances as are seen in Figs. 130 to 132. Or the extension may result in elongation either above or below

tion, are characteristic, and in a family like the Melastomaceae, from which most of the above illustrations are taken, possess generic value. Instead of elongating as a single body, the base may apparently divide longitudinally, through extreme broadening, resembling a forked filament, one theca borne on each branch (Figs. 138 and 139). When one of the thecae then becomes suppressed, its connective branch

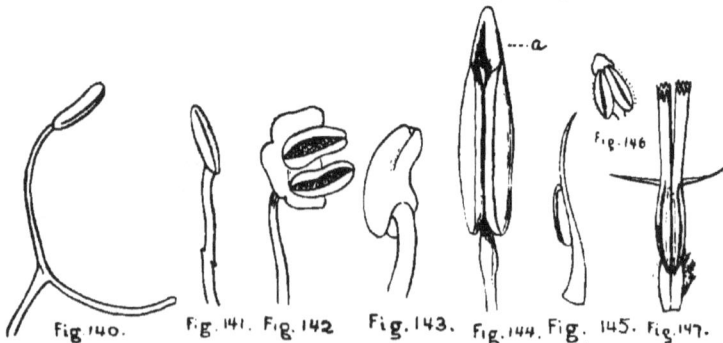

···a

Fig. 146

Fig 140. Fig. 141. Fig. 142 Fig. 143. Fig. 144. Fig. 145. Fig. 147.

the thecae instead of in broadening. If the extension is downward it will lead to an apparent jointing of the filament (Figs. 133 and 134). A slight bulbous enlarge-

remaining (Fig. 140) or even disappearing (Fig. 141), one form of the one-celled anther results. Another form is produced by simple abortion, without any such modification of the connective. or it may result from the disappearance of the connective. Instead of the base, the back of the connective may be appendaged. It may become expanded into a disk-like form over the backs of the thecae, as in Gratiola (Fig. 142). The backs of the anthers may be excavated to receive it, as in Aloe (Fig. 143), or it may be appendaged in any other direction. Appen-

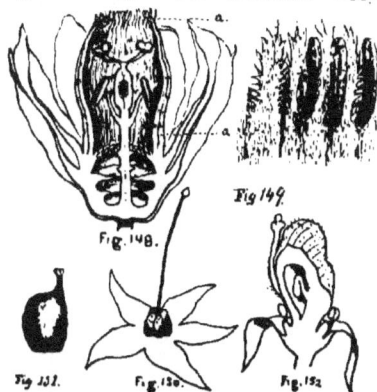

Fig 149.

Fig. 148.

Fig 151.   Fig 120.   Fig. 152

dages of any form may develop at its apex. In the Compositae these are frequently triangular, as in Eupatorium (Fig. 144 a), or lance-shaped. In the Asarum (Fig. 145) it is an awn, while in the Violet (Fig. 146) it is sail-shaped. Sometimes it is formed like a feather (Plumose). It remains to be pointed out that the thecae themselves may be similarly appendaged at any part. Fig. 108 displays Caudae, or tails, which are found in a great variety of forms. In Fig. 126a, Alae, or wings, are illustrated. Fig. 147 displays dorsal spurs or claws (Calcarate) which frequently are borne at the top. Apical awns to the thecae, forked and pore-bearing at the summit, are also shown in Fig. 147. The corona, described in connection with the corolla, is developed to a remarkable degree in the passion-flower (Fig. 148 a). The stamen-column itself is subject to remarkable and characteristic appendaging, with or without connection with an adnate disk. Ordinarily, the summit of the stamen-tube terminates at the beginning of the distinct portion of the stamens, but sometimes, as very generally in the Amaranthaceae (Fig. 149), it is continued up-

ward in the sinuses of the anthers, and this portion may be lobed and appendaged in the most beautiful manner.

Stamens which extend beyond the margin of the corolla are called Exserted or Exsert. This term is also applicable to any organ which projects beyond the perigone.

THE GYNAECIUM.

It has been shown that the gynaecium, except in those rare cases in which a central appendage of the torus is projected upward, occupies the centre or summit of the flower; that it consists of one or more carpels or carpophylls which may be all coherent into a single pistil, the Syncarpous, Gamocarpous or Compound Pistil (Fig. 155), or may each form a separate pistil, the Apocarpous, Monocarpellary or Simple Pistil, and that ordinarily the carpels alternate with the stamens of the adjacent circle. The parts of the pistil have been defined and it has been shown that of these the stipe or thecaphore is rarely present, and that the style is very frequently absent, resulting in the Sessile Stigma. The different forms of adhesion and its effects, as well as those of suppression and metamorphosis, have also been explained. Some additional facts of a general nature must be considered before taking up the details of this subject. The student should from the outset resist the temptation to seek the characters of the gynaecium in the mature or immature fruit, because of its more convenient size. While many of the characters of the gynaecium are permanent, there are many others which entirely disappear after the fertilization of the ovules, and others which only then make their appearance. The other parts of the flower should be completely stripped off, this operation being performed under close and continuous scrutiny, with the idea of detecting any characteristics of relationship between them and the gynaecium. The latter should then be carefully examined in situ. An implement should be passed down between the carpels to determine what degree of cohesion, if any, exists between them, for this will occasionally be found at the very base only, and also to determine if there be any adhesion to a central prolongation of the torus. The details of attachment to the torus must also be determined and their arrangement considered. When numerous, the pistils are apt to assume the spiral arrangement which has already been noticed in referring to position of floral parts. When solitary, the carpel assumes a position to one side of the axis, thus demonstrating its isolation by the suppression of the complementary parts of the circle. A lack of uniformity, as indicating abortion of one or more carpels, must be looked for. When all are uniformly aborted, in the case of flowers which are hermaphrodite but imperfect, this fact will sometimes escape detection unless both forms of flower are examined. The color, texture and surface of the carpels call for minute examination in all cases, though there are no peculiarities of a general nature differing from those of the other organs. As in the case of the petals, so in that of the carpels, the general form is determined by that of the foliage leaves; but the form is less closely preserved and the homology is far less apparent here than there, owing to the far more profound modifications which are rendered necessary by the peculiar functions of the carpels, a consideration which will further on be seen to apply with special force to the fruit condition.

The position of the style often calls for scrutiny. It does not always rise, as would be expected, from the summit of the ovary. One process by which deviation results is illustrated by Fig. 150, which represents the deeply lobed ovary of borage, the single style rising from the fissure in the centre. If, now, all but one of the parts of such an ovary were to become aborted, the style would be seen rising more or less laterally (Figs. 151 and 152), or even basally (Fig. 153), from a monocarpellary ovary. Even though the styles remain separate in such a divided ovary, yet their insertion is necessarily carried toward the base (Fig. 154).

The same descriptive terms already applied to the filament apply equally to the style and its branches. Owing to the frequency with which styles are coherent, ribbed, channelled or angled forms are common. Fig. 155 illustrates the coni-

cal style of Piper, Fig. 156 an obconical one; Fig. 157, one obconico-prismatic; Fig. 158, a clavate form; Fig. 159, one with a bulbous base; the style branches in Fig. 160 are filiform; in Fig. 161 they are filiform and plumose; in Fig. 162, capillary, and in Fig. 163, subulate.

The position and form of the stigma are of very great importance in classification. It has already been shown that while the stigma is commonly located at or near the apex it may extend, either entire or divided into two lines, for a greater

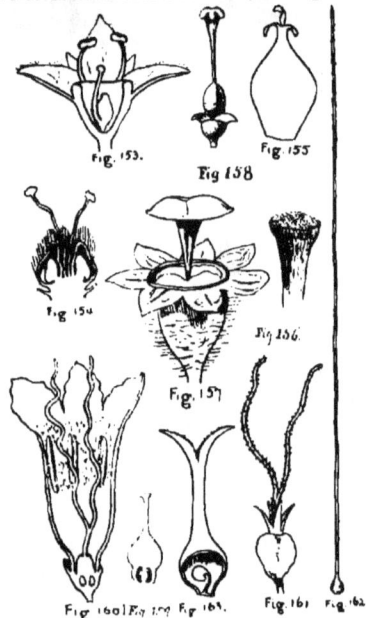

Fig. 153.
Fig. 155
Fig 158
Fig 154
Fig 156.
Fig. 157
Fig 160 Fig 159 Fig 163.
Fig 161
Fig 162.

or less distance down the ventral margin of the style, becoming Linear (Fig. 164). If several united styles are separate at the summit, or upper portions, their stigmas are commonly borne upon their inner faces, as in this case, and are frequently, by the cohesion of the former in the young condition, secluded from the access of pollen until a certain time (Figs. 164 and 236). Between the condition of complete separation and complete cohesion of several stigmas there are all degrees of division and of lobing of the divisions (Figs. 165 to 168). A stigma which is strictly terminal and more or less spherical, thus resembling a head, is Capitate (Fig. 169). A capitate stigma is often Truncate (Fig. 170). If flattened and attached at the centre it is Peltate (Figs. 157 and 171), and this may be horizontal or oblique, as in the latter. The

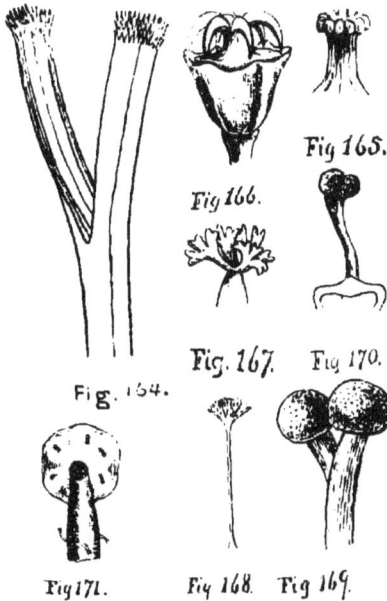

Fig 165.

Fig 166.

Fig. 167.

Fig 170.

Fig. 154.

Fig 171.

Fig 168.

Fig 169.

peltate stigma may have its margin reflexed, making it umbrella-shaped (Fig. 172), or upturned, making it cup-shaped, or Cupulate (Fig. 173), and either of these forms may be lobed (Figs. 174 and 175). Several oblique laminar forms are shown in Figs. 176 to 179, b, in the latter, displaying the manner in which it enfolds the stamen. The

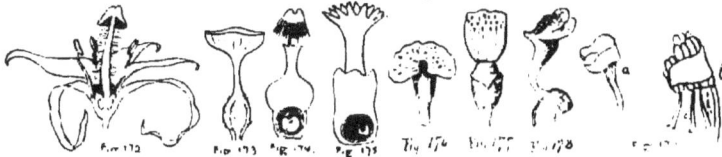

stigmas sometimes form a ring at or below the apex, the Annular form, various modifications of which, unlobed and lobed, are shown in Figs. 180 to 184. Such forms prevail in the order Apocynaceae and are of great value in classification. A stigma (or other organ) with a brush-like plume or appendage is called Penicil-

late (Fig. 187). A Plumose stigma is shown in Fig. 188. Such prevail among the grasses.

The number of carpels in a compound pistil is indicated by the use of the appropriate numeral followed by the suffix "carpellary," thus Dicarpellary, Tricarpellary. The determination of the number of carpels is of the utmost necessity, but is usually a difficult task for the beginner, especially if not previously trained in the art of plant dissection. The indications may be divided into external and internal. The latter must be apprehended from the study of internal structure explained below. The external indications are as follows:—The number of pistils when these are monocarpellary or simple (Figs. 22 and 23); for complete chorisis of a carpel (except apparently in fruit), producing a duplication, never exists. If cohesion is partial, even though so nearly complete as to leave a separation represented by a mere lobing at apex (Fig. 189) or dorsum (Figs. 190 and 191), the determination is scarcely more difficult. It is true that the latter condition is often complicated by grooving or pseudo-lobing pertaining to the backs of the individual carpels, but such lobes are usually characteristically different from those separating the carpels. While the above remarks have been applied especially to the ovary, they may be applied with equal force to the styles and stigmas. If the exterior of the ovary bear no indications of the number of carpels, we may count the styles, or the divisions or apical or dorsal lobes of a style-column, and if all those be wanting, then the corresponding characters of the stigmas or stigma. It must be noted, however, that complete or partial chorisis of style or stigma is not at all rare, and care must be taken to avoid falling into error. In such case the number of lobes of each is apt to equal the number of styles or stigmas. In the case of failure of all these indications to appear, the internal structure must be studied. For this purpose both longitudinal and transverse sections must be made. The former should

be so directed as to lay open the inside of a carpel, and of the latter there should be three, through the lower, middle and upper portions respectively. In most cases a good lens will be sufficient to present the characters, but when insufficient, recourse must be had to the stage and low power of a compound microscope.

Although appendages to the pistil during the flower-stage are less frequent and less varied than in the case of the androecium, yet none of the forms which we have there observed are here excluded.

Two distinct types of the carpel respectively characterize the Gymnosperms and the Angiosperms, both of which classes contribute important medicinal plants. The essential character of the former is illustrated in Fig. 192. This consists in its not being shaped into an enclosure for containing the ovules. In this

instance there is no progress toward such a condition, the carpel remaining more or less flat and bearing the ovules upon its surface. In the progressive forms there is a cavity formed, but it is never completely enclosed. A higher development of it is found in the Taxus or Yew (Fig. 193). The pseudo-cavity of the gymnospermous carpel is never divided.

There are two modes of the enclosure of the cavity of the monocarpellary angiospermous ovary. In the first (Fig. 194) the margins of the carpel meet one another, and then, by more or .ess of an involution, form the placenta (a), with its two rows of ovules (b and c). By the other mode (Figs. 195 and 196) the margins turn in and meet the midrib, forming two cavities, each containing half of the placenta with its one row of ovules. The posterior portions of the carpels thus brought into juxtaposition (a and b) may unite (Fig. 196) or remain distinct (Fig. 195). In other cases where a similar condition exists, it has resulted from the outgrowth from the midrib, across a cavity of the first form, of a false wall or dissepiment, as in the flax (Fig. 197). The placenta and ovules are then found upon the ventral side. It is to be remembered that these are then false cells, each two being indicative of one carpel. With rare exceptions the distinction is to be made by observing the separation of the two rows of ovules. If we imagine two, three or more carpels, constructed in any of these ways, standing face to face, and cohering in this position, we have a perfect idea of the simplest forms of the syncarpous pis-

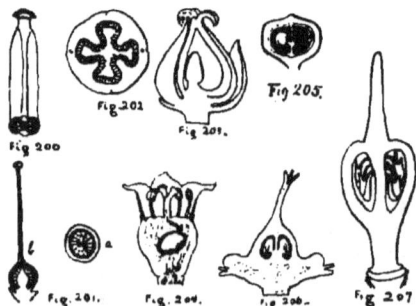

til. In the one case (Fig. 198) we shall have as many walls and cells as there are carpels; in the other, falsely twice as many of each, but the number of rows of ovules should be the same in either case. In the **mustard** and its relatives the false wall connects the two ventral instead of the dorsal sutures. While the lower portions of the carpels are acting thus to form the ovary it must not be forgotten that the upper portions are also uniting to form the styles and stigmas, or stigmas alone, as the case may be. Should the edges remain everted along more or less of the style, they may fail to produce an epidermal covering and become stigmatic, giving us a marginal style. Should they be involute like the ovarian portions, the stigma will be confined to the apex of the style. In the syncarpous form there will merely be a multiplication of these effects.

Quite a different group of appearances will result from the higher or more complex form of carpel union, by which the proximate margins of two adjacent carpels meet and unite (Fig. 199) instead of two belonging to the same carpel. The result of this form must be a single cavity or a one-celled ovary, unless, as in the Mustard (Fig. 200) one or more false septa may divide it. Here, as before, however, it is observed that the ventral portions of the carpels are directed toward the centre of the flower, the dorsal facing outward. This law is invariable, no matter what the number of the carpels or the form of union.

It will be observed that in all cases in which the margins of the same carpel unite with one another, the placentae will be formed at the centre or axis of the flower. Such placentae are therefore called Axile or Axillary (Figs. 198 and 201). Where the margins of different carpels unite (Fig. 199) the placentae must be formed upon the walls or parietes, and are therefore called Parietal. Such placentae may by an extensive involution of the margins be carried very nearly, or quite, to the axis (Fig. 286), but unless cohesion actually occurs at that point they are still parietal and the ovary is one-celled. Some further modifications of the placenta, either in itself or resulting from modifications in the dissepiments, must be carefully considered. If such a pistil as that represented in Fig. 198 shall

develop a condition in which the dissepiments are wanting the placentae will be left unsupported in the axis (Fig. 201), and will then be known collectively as a Free Central Placenta. By progressive abortion of the upper portion this placenta may become reduced to a trace at the base. Upon the other hand, such a placenta may become enlarged and fleshy. Similar changes may occur in the parietal placenta. It may become reduced to a mere point preserved at the apex, base or intermediate portion. In the watermelon it becomes enormously enlarged, filling the entire cavity with a fleshy mass (edible). In the Obolaria (Fig. 202) it is laterally expanded to form a more or less complete false lining to the ovarian cavity. In this position it may remain free or become coherent, so that the entire face of the ovary may appear to be ovuliferous. In other cases it actually is the entire inner face of the carpel, which becomes the placenta. By a subsequent obliteration of a portion of such an expanded placenta the remaining portion may be seen to assume an abnormal position, being occasionally confined to the midrib itself.

As has already been pointed out, the number of ovules is extremely variable and the proportion of them which become fertilized is little less so.

The position of the ovules is to a great extent determined by the nature of the placenta, as has already been explained. It calls for a number of distinctive terms. The two rows of ovules produced by the two carpellary margins do not always appear distinct, but may be reduced, before or after fertilization, to one. A vertical row of ovules is called a series, and ovules are thus defined as being One-serialled, Two-serialled (Fig. 194), etc. When there are many series, so that the number is not readily made out, we simply say that they are Many-serialled (Fig. 202). Ovules placed side by side (Fig. 194), are called Collateral. Sometimes no definite series can be made out, owing to the crowding of many ovules into a small space, as in Obolaria (Fig. 202). They are then said to be Crowded. Collateral ovules, and, indeed, any ovules standing together and deviating from a straight line, have a tendency to turn their foramina away from one another.

As to the directions, in relation to the ovary, which ovules assume, they are Erect (Fig. 203) when standing erect from the base; Suspended (Fig. 204) when occupying an exactly opposite position; Horizontal (Fig. 205) when taking a direction at right angles to the axis of the ovary; Ascending (Fig. 206) when directed obliquely upward from some point intermediate between base and apex, and Pendulous (Fig. 207) when directed obliquely downward from such a point. When starting as an ascending ovule and afterward drooping (Fig. 208) an ovule is Resupinate, or when as in Fig. 209, Recurved-pendulous.

An ovule may have its direction obscured by peculiarities of attachment. Thus, in Loxopterygium (Fig. 152), the real base becomes, by extreme obliquity, apparently lateral and causes an erect ovule to be apparently ascending. That of Anemone is suspended, but owing to the same condition apparently only pendulous. The terms erect and suspended are after all only relative, as we can never be sure that an ovule which appears in such position is really the uppermost or lowermost of its series. Very often others which would have been in reality the basal or apical have become aborted, as in the last case illustrated.

A merely recurved ovule is not to be mistaken for an anatropous ovule. The latter has the contiguous portion of the funicle adherent as a rhaphe, which comes away with the seed at maturity.

The recognized varieties of ovules are based upon the external structures, which will here be briefly considered. The details of their inner structure will be considered in our chapter on fertilization. The ovule consists of a Body (Fig. 209, a) and Funiculus or Stem (b). Named in the order of time in which they are developed the parts of the body are as follows: The Nucellus, or central portion (Figs. 210-213, n), containing the parts essential to reproduction, and two coats, the Primine or inner (k) and Secundine or outer (s). Certain parts of these, or points upon them, also have distinctive names. The more or less circular opening (m) left

at the apex by the failure of the coats to completely inclose the nucellus is the Micropyle. The structurally opposite end of the body, or the point where nucellus,

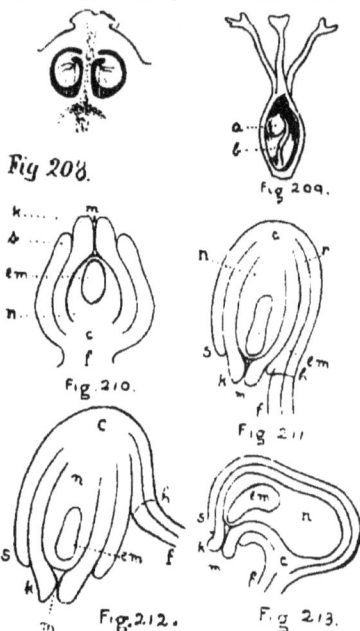

Fig 208.

Fig 209.

Fig. 210.

Fig 211

Fig. 212.

Fig 213.

coats and apex of funiculus separate from one another (c), is the Chalaza. If the body become inverted upon its funiculus, either partly (Fig. 212) or wholly (Fig. 211), the portion of the funiculus against which it lies (r) will become adnate to it and is known as the Rhaphe (also spelled Raphe). The portion of the funiculus remaining free (f) is then specifically known as the funiculus. When hereafter in this work the last term is used it will be understood as applying to this free portion. It is thus seen that the rhaphe is limited at its distal end by the chalaza; but separation of this seed at maturity cannot take place at this point, owing to the adnation of the rhaphe, as it would do if no such adnation existed. Separation in such case must take place at the point where rhaphe and funiculus join; hence the Hilum, as such point of separation is called, may be variously situated, and need not coincide with the chalaza. In Fig. 210 it is at the chalaza, in Fig. 211

at the opposite end (h), while in Fig. 212 (h) it is about half way between. The parts here enumerated are not always conspicuous and may be easily overlooked by the beginner.

The nucellus is the essential part of the ovule, which in some cases consists of nothing else, and even this may be reduced to its lowest essential elements. An ovule without either coat is Naked or Achlamydeous; with only primine it is Monochlamydeous, and with both it is Dichlamydeous. An ovule without funiculus, and the same is true of any organ not borne upon a stem, is Sessile. The form of the funiculus, as well as its direction, always calls for inspection. It may be very short and broad (Fig. 210), or elongated and slender (Fig. 209), and the latter form may be either straight or variously curved. An ovule with a rhaphe is Anatropous when completely inverted (Fig. 211), the rhaphe running its entire length; Amphitropous when this condition is but partial (Fig. 212). When the body of the ovule is doubled (Fig. 213), its relation to the funiculus not considered, it is Campylotropous. An ovule which is none of these, being both straight and erect, is Atropous or Orthotropous (Fig. 210).

Before proceeding to the subject of pollination and fertilization, and the changes in the several parts of the flower consequent thereon, we must consider in detail the torus and its modifications.

THE TORUS.

The fundamental principles of anthology are based upon the nature of the torus as a modified branch. We have already considered the evidences of this fact depending upon its position and the relative positions of the parts developing upon it. We shall now consider some which depend upon its modifications. These are in part permanent and typical and in part exceptional and abnormal. Among the latter we note that in those frequent cases in which the parts of flowers revert to the leaf condition the torus often elongates, separating the floral series exactly as whorls or spirals of leaves are separated upon a branch. At other times it will be continued beyond the apex or centre of a flower in the form of a leafy branch. Occasionally one of the sepals will be found at its proper radial point, but distant from the rest of the calyx, a portion of the flower stem intervening. A similar condition, but affecting an entire series, normally characterizes certain species or groups of species. The elongation may affect any internode or internodes. When (Fig. 214 a) it is between

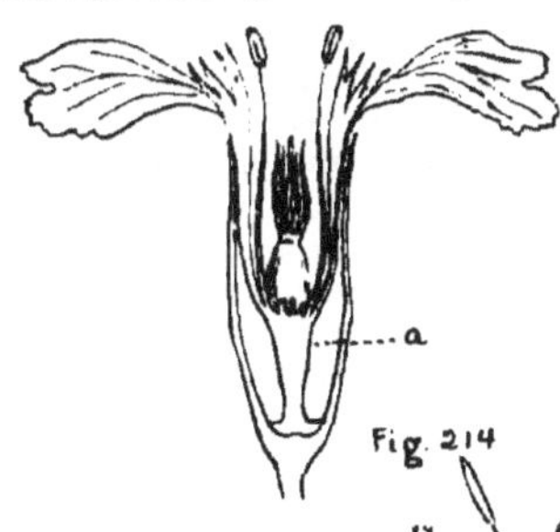

Fig. 214

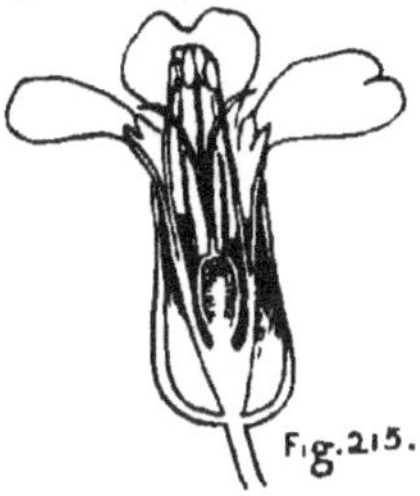

Fig. 215.

Fig. 216.

Fig. 217.

Fig. 218.

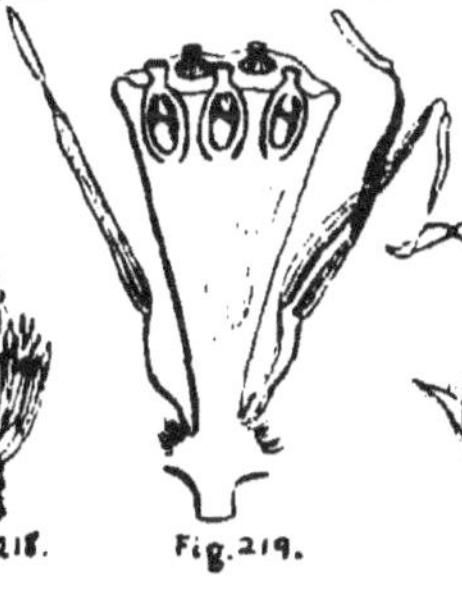

Fig. 219.

Fig. 221.

Fig. 220.

calyx and corolla it is called an Antho-phore. Sometimes, as in Viscaria (Fig. 215) the anthophore may be very slight, so as to escape detection until a longi-tudinal section reveals its presence. A similar elongated portion between corolla and androecium is a Gonophore (Fig. 216, a). One between androecium and gynae-cium (Fig. 216, b) is a Gynophore. A thecaphore (Fig. 9) often resembles a gynophore and may be mistaken for it. The point of articulation and separation at maturity will determine whether the stalk is a portion of the ovary or of the torus. A slender extension of the torus upward among the carpels, which are at-tached to it, constitutes the Carpophore, as in Geranium (Fig. 230). The presence of a carpophore is almost characteristic of plants in the Umbelliferae (Fig. 230A). In the Boraginaceae it is frequently re-duced to a pyramidal or conical form, or shortened and laterally expanded until it it is merely convex or even plane. To all such modifications the term Gynobase is applied. In this condition it may become hollowed out at the insertion of the car-pels, as in borage (Fig. 217). In all forms of the gynobase it is important to note the point of attachment of the divisions of the ovary and the scars which the latter leave upon removal. The above consid-erations refer to elongations of internodes of the torus. The condition of adnation of floral parts may, upon the other hand, be looked upon as one in which the normally very short internodes of the torus are still further shortened, so as to bring the parts into most intimate connection. In-stead of undergoing a mere elongation of its internodes the torus may be laterally expanded at any or all points, with or without elongation also, and in innumer-able forms. An expansion or appendage of this kind, although the term may prop-erly be regarded as including all forms of enlargement or expansion of the torus, is called a Disk. The simplest form is perhaps that seen in the blackberry (Fig. 363, although the most of the enlarge-ment here seen, as in the next, is the ac-crescence of fructification), that of a hem-isphere with the pistils arranged upon its surface. The disk of the strawberry (Fig. 262) is similar, but the pistils are partly immersed. In the rose, a related plant (Fig. 46), the form is modified by the elevation of the margins instead of the centre, so that a cup-shaped disk is formed, the pistils attached over its inner surface. In the cherry (Fig. 44), also related, and the apple the disk is thin and lines the calyx tube, the one (in the former) or few (in the latter) pistils being centrally placed, and in the former re-maining free from the disk. In the mag-nolia (Fig. 218) the torus is vertically much elongated and at the same time much thickened, the pistils adnate along its surface. In the Nelumbo the torus (Fig. 219) is enlarged into a top-shaped or Turbinate body, with the pistils im-bedded in the flat upper surface. Instead of thus occupying a hypogynous position the disk may be projected between any two of the circles, and it may be wholly or partly adnate to either (Fig. 228a), or both of them or it may be entirely free. When adnate to both it is plain that it becomes responsible for the existing ad-nation between the latter. It may then exist only at the base (Fig. 228), or it may entirely fill up the interspace between the parts and even become epigynous, so that the ovary is immersed in it or buried underneath it (Fig. 220a). The adnate disk may be shorter or longer than the circle to which it is adnate. The simplest manifestation of the disk is that of a mere swelling or ring (Fig. 221) at the summit of the torus; its greatest that in which it becomes an elongated cup or tube (Fig. 222). Either form may be en-tire or more or less divided, from that with a mere sinuately lobed margin (Fig. 223) through the toothed and lobed (Fig. 224) to that consisting of entirely sep-arate divisions (Fig. 225). It may be reg-ular, as in the above illustrations, or very irregular (Fig. 226), and cohesion may exist between some of its divisions while the others are free (Fig. 227). The lower portion may be adherent while the upper, lobed or entire, will be free (Fig. 228). It may be itself appendaged, and it may or may not be glandular in nature. Fi-nally, we note that the disk may be double, its two circles occupying different internodes of the torus. The texture of the disk is commonly thicker than that of the other parts, but it may be laminar. It is therefore sometimes easy to mistake a disk for a corolla, aborted stamen circle or crown. In all its peculiarities above described, and in the number, size and form of its divisions and appendages, the disk is characteristic and of the greatest value in classification, either generic, as in the Gesneriaceae, or specific, as in Eschscholtzia.

POLLINATION.

We have seen that the essential female element of reproduction in the flower is produced in the nucellus of the ovary, the male within the pollen-tube. We have also seen that these two elements are

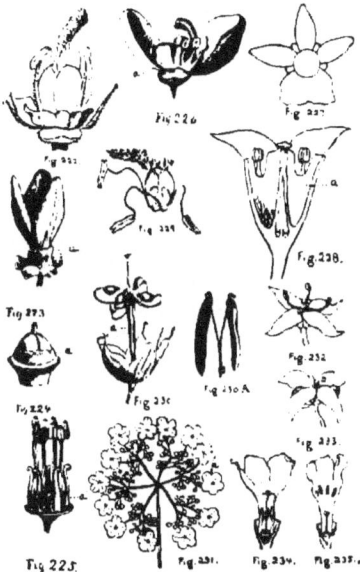

produced separately, and in most cases remotely, from one another, and that some means must exist for bringing them together in order that fertilization may be effected. In those plants (Gymnosperms, Figs. 192 and 193) in which no stigma exists, this is accomplished by bringing the pollen into immediate contact with the ovule, which is exposed to external contact. In those in which a stigma exists it is accomplished by the deposit and fixation of the pollen thereupon. To either of these processes the term Pollination is applied. The two elements may proceed from the same flower, in which

case the term Self-Pollination or Close-Pollination is applied, or they may proceed from different flowers, in which case the term Cross-Pollination is applied. It will be noted further that there are degrees of cross-pollination, according to whether the elements proceed from flowers upon the same or upon different plants. When the flowers are perfect it is at least possible in most cases for them to be either close or cross pollinated. In nearly all cases the reproductive function is strengthened through cross-pollination, which explains the fact that nearly all flowers are constructed so as to facilitate cross-pollination, while most of them are so constructed as to incommode, and very many to prevent, close-pollination. In a few cases the flower is constructed so as to prevent cross-pollination. The methods of effecting pollination may be divided into the ordinary and the exceptional. The latter must be considered individually. The former are two—namely, through the agency of the wind and through that of insects (or occasionally other animals). Flowers adapted to the former method are called Anemophilous; those adapted to the latter are called Entomophilous. Occasionally the flower is so formed that the movement of the water during rains, or in streams, effects pollination. The activity of the wind being beyond the control of the flower, the adaptation of the structure of an anemophilous flower is limited to securing the benefits of such action when it comes into play. This consists chiefly in (1) a Gregarious Habit—the growing together in great numbers of individuals of one kind, as in the case of grasses and of most of the forest trees of temperate latitudes; (2) a very abundant pollen (3), which is loosely fixed (Fig. 229, one method), light and easily removed and transported, and (4) the disposition of the ovule of gymnosperms, and the form and disposition of the stigma and connected parts of angiosperms, so as to catch the pollen. All these provisions may be readily seen to affect the process in the case of Pinus palustris, for example.

In entomophilous flowers such provisions must be preceded by others of a different nature, calculated to attract and excite the action of the forces to be utilized by the former. It is the possession of both of these classes of provisions which constitutes one of the most important distinctions between flowering plants and the flowerless, in which latter the male element is almost invariably provided with the power of independent locomotion, by which it can reach the female. Provisions for attracting external agents are found chiefly in the form, coloration and size of the flower or of one or more of its parts, the production of fragrant and nutritive secretions and the exercise of these influences at the most opportune times. The form of the flower is efficient when it resembles a form attractive to an insect the visit of which is desirable, or when it is one well calculated to display effectively the coloration; and it is not impossible that certain forms, like certain colors, are attractive *per se*. The forms of nectar-bearing plants are, moreover, in most cases, such as to facilitate the collection of the food by the visiting insect, or, when otherwise, to effect special objects to be considered further on. Coloration also may be attractive through its simulation of an insect or merely by its serving to make known to the insect the presence or position of the flower concerned—as a white, light-colored or lustrous flower, in attracting insects which fly only when there is little light. Flowers are frequently modified in size so as to effect these results, and this modification is often secured at the expense of their own sexual functions. Fig. 231 illustrates a cluster of Viburnum flowers, the marginal being light-colored and large, and admirably adapted to attract insects, but destitute of perfect reproduction parts. The odors of flowers similarly, while frequently offensive to the human sense, are supposed to be attractive in most cases to the insects whose visits favor their pollination. They are due to the evaporation of volatile oils. The glands by which these are excreted and in which they are stored may be distributed through the tissues of all or certain of the floral parts, or their presence may be restricted to the special appendages described below. The nutritive substances other than pollen to

be consumed by the visiting insect, known as Nectars, are produced by certain special glands and are stored in or upon contiguous receptacles called Nectaries. The presence of these nectaries is commonly responsible for the outgrowth of the appendages to which they are often attached (Figs. 52 and 53). At other times a part of the flower not conspicuously modified produces and holds the nectar.

The influences here described are in almost all cases exerted at certain times which are especially favorable for securing the desired results. In speaking of the perigone it has been shown that flowers vary greatly as to their duration. It may be further stated that those which perish quickly mature and expand at the particular time of day when pollination is most likely to occur. Those which last for several days enjoy a daily resting period and another period of greatest activity, the details of which vary in different species or classes. Commonly the perigone becomes more or less folded or closed, its form and coloration less conspicuous, the exhalation of odors entirely suspended or greatly restricted and access to the nectar prevented altogether. At the same time that its functions are thus inactive its position is such as to afford it protection of various kinds from dangers which are especially imminent during the hours in which it rests. This condition of inactivity or rest is commonly spoken of as the sleep of the flower. It occurs at such a period of the day as finds the agencies specially adapted to pollination in its case themselves enjoying their rest. As these again become active, the flower "awakens" and all the conditions above noted are reversed, or at least such of them as affect the flower in question. Flowers in which this active period occurs during the day, whether they endure for but one day or longer are called Diurnal; those in which it occurs at night are called Nocturnal. Besides the regular daily resting period, a great many flowers, by virtue of special sensitiveness, possess the power of assuming such a condition on special occasions when the conditions call for it.

Humming-birds, as well as insects, are

active participators in the operations above recorded. Their operations in promoting cross-pollination in the Cinchona group have been largely responsible for some of the most far-reaching economic conditions and results in the history of the drug trade. In exceptional instances still other animals take part in this work.

It may be remarked in passing that these characters, like some of those which follow, are not restricted to the flower itself. Very frequently other portions of the plant adjacent to the flower will be expanded, brightly colored and developed into special forms, while the odor of some flowers, due to the presence of glandular tissues, is shared by the foliage and other herbaceous portions, as in the lavender. Well formed, large glands are present in

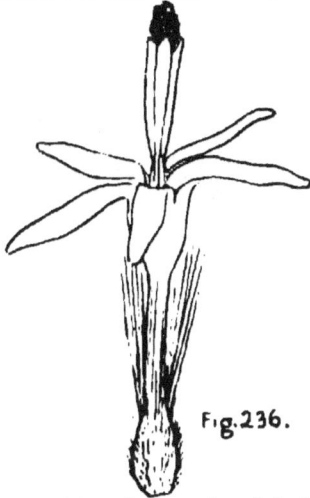

Fig.236.

the axils of the primary veins of the leaves of some species of Cinchona, although the precise function which they perform is by no means clearly established.

The special contrivances for utilizing insect-visits in effecting pollination are far more elaborate and varied than those for inducing them, which we have already considered, and our consideration of them cannot be extended beyond what is necessary to indicate their general nature end classification, and to serve as a key in understanding the complicated modifications which we have observed the typical flower to undergo. Usually the effects extend in two directions: (a) toward excluding the

pollen from access to the stigma of its own flower, and (b) toward securing its access to that of another. One of the most frequent methods of securing the former result is the maturing of the androecium and gynaecium at different times. This method is called Dichogamy. By it the ovules of a flower are already fertilized before the mature pollen of that flower escapes from its thecae (Proterogyny), or else the pollen is matured and' utilized before the stigmas of that flower are prepared for its reception (Proterandry, Figs. 232 and 233). Dichogamy is very common among perfect anemophilous flowers, where self-pollination would otherwise commonly result, and it may be assumed to have been the first step toward the uni-sexual state, so common among flowers of that class. Careful notice should be taken of the fact that in dichogamy the retarded state observed in androecium or gynaecium is but temporary, and that the finally developed forms of both the proterandrous and proterogynous flowers are practically the same.

A far more profound modification is that in which there is a permanent change in the androecium (Fig. 234) of one flower and a similar change in the gynaecium (Fig. 235) of another, by which a similar result is obtained to that proceeding from dichogamy. Such a provision constitutes Dimorphism. By a modification of it a third form of flower, intermediate between the other two, is produced, constituting Trimorphism. As will be seen by a consideration of the following typical examples of each, dimorphism is more intimately connected with the transferring of the pollen than is dichogamy, though the latter is rarely without some special provision for thus supplementing the effect which it produces in excluding the pollen from the stigma of its own flower.

Fig. 236 illustrates a flower of Vernonia. Its anthers are closely syngenesious and introrsely dehiscent. Its style is two-cleft, the stigmas existing upon the inner faces of the branches. It is obvious that until these branches separate pollination cannot take place. Before such separation occurs the tip of the style is, by elongation, slowly forced up through the tube of the anthers. The latter, with the contained pollen, are mature, and the pri-

len is, by the stiff hairs upon the backs of
the style branches, torn out from its receptacles and exposed to such agencies of
transportation as may be prepared to act
upon it. Cases are even known in which
the tearing out of the pollen in this way
is effected by a spasmodic shortening of
the stamens upon the instant of contact
by a visiting insect, the pollen being by
the same process at once discharged upon
the body of the latter. After the removal
of the pollen, or after the death of such
grains as fail to be removed, the style
branches separate in readiness to receive
the pollen brought from some other flower.
This method, or some modification of it, is
very common among the Compositae, and
illustrates how the study of pollination
serves to explain many modifications of
flower-structure otherwise inexplicable,
and why the possession of the latter is regarded by the biologist as indicating a
higher stage of development. The explanation of the case of dimorphism exhibited in Figs. 234 and 235 is as follows:—An insect visiting flower No. 1
and thrusting his proboscis deeply into
the corolla tube in search of nectar brings
his body into contact with the stamens,
and pollen is deposited upon it. The next
flower visited may be one of the same kind
or one with the long style. If the latter,
then the portion of the body which is now
covered with the pollen is brought into
contact with the stigma, upon which the
pollen is deposited. At the same time a different part of the body is being laden with
pollen from the short stamens of flower
No. 2, to be deposited upon the short pistil
of still another flower, similar to No. 1.
If perchance two flowers of the same
form are visited in succession the result
is that an additional deposit of pollen is
secured, or at most a portion of the pollen
already being carried is left upon the
stamens of the visited flower. In conclusion it may be said that even if, by some
failure in the provisions here described, the
flower should become self-pollinated, we
have excellent reasons for believing that
pollen from a different flower which might
be deposited at the same time would find an
advantage accorded to it by which it
would be enabled to first reach and fertilize the ovules.

The assuming of a form convenient for
the visiting insect, to which reference has
been made, is very frequently interfered
with for the purpose of forcing the insect
into such a position as shall favor or com-
pel the removal of the pollen, a labor
which is by no means agreeable to it and
which it not rarely seeks to avoid, as in
the case of the bee, which cuts a hole at
the base of some corollas.

In spite of the possibility of thus effecting a rough classification of some of the
methods of securing cross-pollination, it
is yet true that the great majority of instances are not subject to classification
and must be denominated special, or else
that they combine some special arrangements with such general methods as have
been described.

Flowers which are self-fertilized before
expansion are Cleistogamous. Occasionally fertilization takes place without the
removal of the pollen from the anther.

Fig. 237.

The pollen thus transferred to the stigma must be fixed there in order that fertilization may follow pollination. This
process is effected by contrivances little
less elaborate, although more minute, than
those which have been described. These
contrivances relate in part to peculiarities
of the pollen which will be found described in the part relating to histology.
As regards the stigma, fixation is effected
most generally by means of the viscid secretion to which reference has been made.
Appendages in the form of hairs, scales,
or papillae (Fig. 237) are very common.
In some cases the divisions of the stigma
are sensitive and close elastically upon
the pollen as soon as it is deposited. With
the fixation of the pollen upon the stigma
pollination is completed and fertilization
begins.

FERTILIZATION.

A knowledge of fertilization is of importance to the pharmacognosist only as it throws light upon the characters of the fruit, in which term we include the seed. Only the principal facts connected with the subject will, therefore, be here considered.

The gross appearance and parts of the ovule have already been described. Its internal structure is illustrated in Fig. 238.

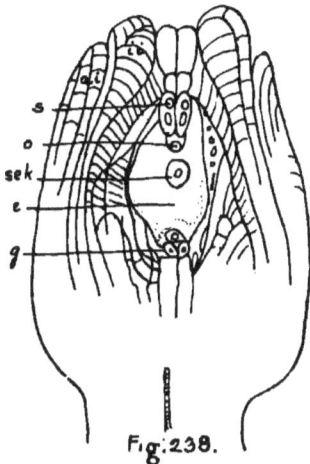

Fig. 238.

ii, primine; ai, secundine; S, synergidae; O, oosphere; SeK, nucleus of embryo-sac; e, embryo-sac; g, antipodal cells.

The Nucellus, being a cellular body, in which one or more spores are to develop, is a Sporangium. The development of its solitary spore presents striking differences from that in flowerless plants, the most important result being that the spore produced is single, and, germinating, fills the entire interior of the nucellus, the cellular tissue surrounding it at the time of its origin mostly disappearing subsequently to give place to it. Almost immediately after its formation the macrospore germinates by the division of its nucleus. Although the development is very simple and the growth very slight during the first

stage, the resulting structure confined entirely within the embryo-sac, it is to be regarded as a plant body, the female gametophyte. It develops seven incompletely formed cells, or corpuscles, possessing specialized functions. That occupying a central position is the Central Cell or Nucleus of the Embryo-Sac. The two nearest and in contact with the foramen are the Synergidae, in contact with which lies one called the Egg-Cell, or Oosphere. Those at the opposite end of the embryo-sac are called the Antipodal Cells. _he mouth of the foramen affords a means of ingress to the fertilizing element in this, the angiospermous ovule. The ovule of gymnosperms agrees in the possession of an embryo-sac, with several bodies corresponding to the oosphere of angiosperms, but the other corpuscles not clearly developed. The foramen is secretory, so as to be adapted to acting upon the pollen-grain which it receives, as does the stigma in angiosperms.

Between the ovule thus prepared and the stigma there is an almost continuous connection through conducting tissue, extending through the body of the stigma, style and placenta. The extent of this conducting tissue, like that of the stigmatic surface, is usually greater or less according to whether there are more or fewer ovules to be impregnated.

The pollen grain consists of a highly hygroscopic mass of tissue, partly vital and partly nutritive, the latter of variable composition, surrounded by a thin, non-perforated, highly elastic membrane, the Intine, and this in turn by a thicker, non-elastic covering, the Extine, or, according to some writers, "Exine," bearing one of more complete perforations, very thin places or otherwise modified points upon its surface. In exceptional cases the pollen grain possesses but a single wall.

The process of fertilization is illustrated by Fig. 239, and the ordinary phenomena are as follows:—The pollen grain, fixed upon the stigma of the angiosperm, or upon the summit of the ovule of the gymnosperm, the hygroscopic contents absorb moisture from the secreting or transuding surface with which it is in contact, oxidation and nutrition commence, the mass increases in size, and distends the intine which surrounds it. At the same time cell-division of its contents takes place, the combined changes constituting the germination of the microspore. Through one or more of the perforations of the extine already existing, or forcibly made by this process, protrude prolongations of the pollen contents, still enveloped in a process of the intine. These processes, or this process, for in general there is but one, penetrates the tissue of the stigma and extends downward through the conducting tissue of stigma, style and placenta, and is known as a Pollen-Tube. A body of this kind, proceeding from the germination of a matured spore, is properly to be regarded, like its female homologue, as a plant body, distinct from that of the parent   It is to be noted that it can be equally produced by germination upon other surfaces which present the proper conditions. It is known as a Male Prothallium, and because, like the product

within the embryo-sac, it is designed for the production of true sexual elements, to result in the formation not of a spore, but of an embryo, it is called in contradistinction to the sporophyte, or spore producing body, the Gametophyte. Nourishment is afforded partly by the contents of the pollen-mass and partly by the tissues with which the prothallium is in contact. At its lower end are one or more little bodies which constitute the male element and which are to fertilize the oosphere, which we have already observed within the embryo-sac. This fertilizing element is the Male Cell, corresponding to the Antherozoid of cryptogams. The downward growth of the male prothallium toward the ovule is known as the Descent of the Pollen-tube. The distal end of the pollen-tube at length enters the ovarian cavity and finds the foramen of the ovule, contact of the male cell with the oosphere is effected and fertilization is completed.

CARPOLOGY.

Fructification or the Changes Produced by Fertilization.

The changes effected by fertilization extend to all parts of the flower and even to other parts of the plant. A consideration of the objects of the process will prepare us to understand the nature of the changes. The objects are (1) the production and maturing of one or more seeds, including provisions for their protection and nourishment throughout the process, together with the nourishment of the parts which thus protect them; (2) provisions for their transfer, still enclosed in their container, to a suitable place for germination and the fixation of the latter there, or (3) provisions for their exit from such container and (4) their transfer after such exit to the place of germination and their fixation there.   The combined processes connected with the attainment of those objects is Fructification, and the product thereof is the Fruit.

It is clear that the energies of the plant should not be called for in the further development or preservation of any parts of the flower which are not serviceable as a part of the fruit in the attainment of the above named objects, unless possibly they may possess some other function foreign thereto, as, for instance, the action of the

stamens of a flower in which fructification has already begun, in fertilizing the ovules of some other flower. We should therefore look (a) for the disappearance or death of all floral parts not thus serviceable, and (b) for the stimulation and development of those which are. That the first of these two objects is an immediate result of fertilization is strikingly and unhappily illustrated in the behavior of ornamental flowers, in which the latter process is allowed to take place. Those who produce for the market the handsome and expensive flowers of orchids are obliged to carefully exclude insects from their greenhouses.   Valuable flowers which, without fertilization, would last for several weeks, wither and die within a few days, or even hours, after such process has occurred. That the accomplishment of the second named object is no less immediate is apparent upon considering the morphology of the fruit.

The only portion of the flower which is certain to be in no case utilized in fructification, and, therefore, to disappear after fertilization, is the actual stigma, and the stamens when they are non-adherent. The stamens, as has been shown, may be serviceable for other purposes, so that their death depends rather upon the performance of their individual function than upon fertilization. In proterogynous flowers this function is actually stimulated by the completion of fertilization in their own flower.

Upon the other hand, we are not certain of a requisition in every case for the preservation and development of any part other than the particular ovules which become fertilized, the ovarian walls of the pistil or pistils containing them (and in

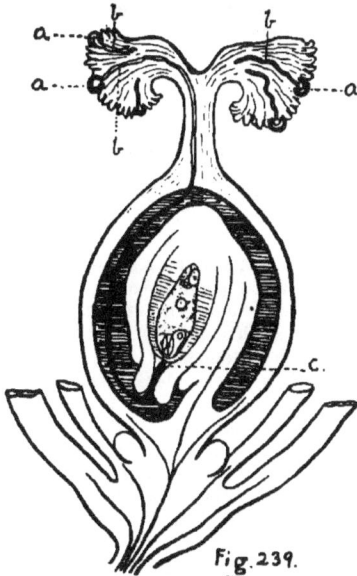

Fig. 239.

a, pollen-grains ; b, pollen-
tubes grown to different
lengths; c, pollen-tube
fertilizing ovule.

some cases only a part of these), and of the torus. The death or decay, therefore, of any or all of the other parts will be determined by the individual or class habit of the plant concerned. To any part other than the ovary itself which thus develops and enlarges as a part of the fruit, the term Accrescent is applied. Fruits of which such accrescent parts form the conspicuous portion are called Accessory fruits.

Finally we must note that new parts, of service in the fruit, frequently develop in the course of fructification, upon either pericarp or seeds, just as special appendages develop upon the floral organs for performing special functions in connection with pollination. That such additional parts exhibit little, if any, development during the floral stage, is due to the fact that an enormous waste of energy on the part of the plant would thus be involved. Of all the flowers produced by a plant only a minor portion usually accomplish fructification, and of all the ovules produced by any gynaecium only a minor portion produce seeds. The development of these superfluous flowers and ovules constitutes in itself a serious waste, but it is a necessary or, upon the whole, an economical one, as it tends in the end to secure the full degree of fructification by the plant. The development, however, upon such superfluous flowers or ovules, of parts which will be of value only in case fructification is effected, would be anything but economical. Hence, the general rule that parts of the fruit which are of no use in effecting pollination and fertilization are not developed until after these functions are performed.

There are two distinct senses in which the term "fruit" may be employed. In the first instance, we may regard it as the structural product of the development in fructification of a pistil, or in the second as an organ performing a certain reproductive function or functions. The limitations of our definition of the term will vary accordingly.

In many cases the ripened gynaecium performs or may perform the fruit function entire, as in the cherry, the strawberry, the blueberry, the so-called "seed" of the sunflower, and the pod of the bean or digitalis. In such cases the solitary ripened carpel (cherry and bean) or the aggregation of ripened carpels (as in the other illustrations), of a gynaecium, constitutes the fruit, from either point of view. In other cases a number of carpels of a gynoecium are separate from first to last as pistils, as in the case of the buttercup. The entire collection then constitutes a fruit, being the product of a flower, but each of the individual pistils must also, from a physiological standpoint, be regarded as a fruit, inasmuch as it performs the fruit function independently.

Again we find, as in the case of the borage, that carpels, originally coherent, separate before performing their function, so that we must regard each of the separated carpels, as well as the entire gynaecium, as in the nature of a fruit. Occasionally even a carpel will itself divide into separate parts, each of which is equally entitled to be designated as a fruit. In still other cases the ripened gynaceia of more than one flower cohere and perform the fruit function as one body, as in the case of the partridge-berry and the fig. Finally we note that many fruits can perform their function in either way—namely, by means of their carpels, or parts thereof individually, or as aggregations proceeding from a single flower (blackberry), or from many flowers (figs, hop, &c.). It is therefore to be noted that that which is at one time to be regarded as a fruit, is at another time only a part of one, according to the manner in which it performs its function.

From the foregoing considerations we may deduce the following definitions of fruit:—

A Fruit is a separate ripened carpel, or part thereof, or an aggregation of ripened carpels, together with any adherent parts.

Multiple or Collective Fruits are those proceeding from the gynaecia of more than one flower.

Aggregate Fruits are those which proceed from a number of separate pistils of one flower.

Simple Fruits are those proceeding from a single pistil.

Apocarpous Fruits are those consisting of one carpel or more than one un-united carpels.

Syncarpous Fruits are those consisting of coherent carpels.

Accessory Fruits are those in which some part other than the ripened ovary constitutes the conspicuous portion.

The student cannot have failed to note in reading the above statements that the composition of the fruit is extremely variable and in some cases complicated. In accordance with this fact the classification of parts of the fruit is open to great differences, according to the principles upon which the observer bases his classification. The typical fruit may be considered as that which consists only of the ripened

pistil with the contained seed or seeds. Such a fruit is regarded as possessing but two portions, namely, the seeds and the Pericarp. But since in many cases the calyx, disk or other part is closely adnate to the wall of the ovary and more or less indistinguishable from it, it becomes impracticable to restrict the term pericarp to a part consisting only of the pistil. Again we find that there are all intermediate forms and degrees of adnation and separation between the ovary and the accrescent parts of accessory fruits. It, therefore, appears most convenient to define the pericarp in a broad sense as the fruit with the exception of the seeds. When the pericarp consists of other elements than the ovarian-wall it is called a Pseudocarp or Anthocarp.

When the pericarp is seen to consist of three demonstrable layers these are called respectively Exocarp, the outer; Endocarp, the inner, and Mesocarp, the middle. When an exocarp is thin and membranaceous, like the skin of a plum, apple, or tomato, it is called an Epicarp, and when an endocarp is hard and strong, like the stone of a peach or the "core" of an apple, it is called a Putamen.

We shall now consider the manner in which the four objects of fructification are accomplished through the modifications effected in each of the floral parts and in the parts adjacent, by fertilization, including such new appendages as are thus caused to develop. The development and maturity of the fruit is effected by the stimulation, through fertilization, of the nutritive functions of the pistil, the torus, adjacent portions of the plant, and through the combined influence of all the flowers, a similar stimulation of all portions of the plant. So far as the development of a protecting container for the maturing seed is concerned, the object in general demands the development of nothing more than the ovarian wall; but the effects of adnation and the requirements of the other objects result in the extension of this process to various other parts of the flower or even of its supporting parts. The development of such parts in connection with the ovarian walls will therefore receive attention in considering the methods by which such other objects are accomplished. It has been stated that not always are all of the ovarian walls involved in the development. A gynaecium possessing several pistils may fail to develop all of them in fruit (Fig. 185), and when these are adnate into a compound ovary, as in Vahesia, one or more of them may likewise fail to develop. A several-celled ovary, as in Calesium (Fig. 239a),

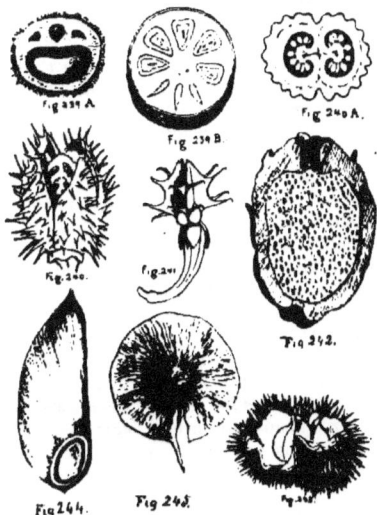

Fig. 239 A.　Fig. 240 A.
Fig. 239 B.
Fig. 240.　Fig. 241.
Fig. 242.
Fig. 244.　Fig. 243.

may, after the fertilization of one or more ovules in one or more cells, permit the abortion of those in the other cells, the septa of the latter being then crowded against the outer wall by the growing seeds, or even disappearing, so that the fruit will contain a smaller number of cells than the ovary which produced it. The partial obliteration of cells in a similar manner is well shown in the fruit of Diospyros (Fig. 239b). Additional walls upon the other hand, may develop during fructification. Datura has a 2-celled ovary (Fig. 240a), but a 4-celled fruit (Fig. 240), and this occurs regularly in the Labiatae. The newly formed walls are not always vertical. The fruit of Desmodium (Fig. 301a) and that of Sophora (Fig. 301) divide transversely into one-seeded joints.

Concerning the protection of the fruit and seeds we note that its full accomplishment often calls for other defensive provisions than those against merely mechanical forces, in the form of appendages constituting an armor. These are sometimes

an outgrowth from the ovary itself, as in Stramonium (Fig. 240), sometimes upon an enclosing calyx (Fig. 241), an enclosing wall, consisting of a hollowed branch, as in the prickly pear (Fig. 242), or sometimes upon an enclosing involucre, as in the chestnut burr (Fig. 243). At other times the protection is secured by developing acrid or otherwise disagreeable pericarps, as the husk of the walnut or the pulp of the colocynth. These defences may be effective only during the maturing stage, as already pointed out, or their deterrent action may be permanent. In the same direction are to be considered the effect of poisonous principles proper and the inedible nature of the pericarp pending the maturing of the seeds.

The transfer of the fruit to the place of germination is secured by methods which for the most part admit of classification. We shall first consider those provisions which utilize the agency of the wind for this purpose.

We note, first, that the weight of fruits to be thus transported is reduced to a minimum. They are in almost all cases one-seeded (Monospermous), the loss due to this character being made good by the fructification of a large number of flowers. The one-seeded condition of such fruits is,

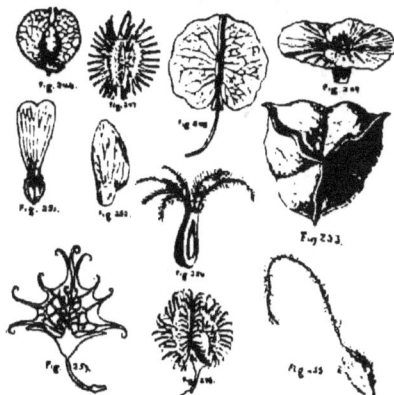

moreover, not restricted to families which are characterized by it. Many fruits of the Leguminosae, which are commonly several or many-seeded, as the pea and bean, become one-seeded when adapted to wind transportation (Figs. 244 and 245). Fruits which are not one-seeded may divide into one-seeded parts, easily separable, to facilitate transportation by wind or other agencies, as has already been shown.

Such a state having been attained, the action of the wind upon them is next secured through the development of an expanded surface of some kind, commonly a wing or plume. In the Machaerium (Fig. 244) it is the entire wall of the ovary, in its original nature a pod, like that of the bean, which becomes developed into a wing. In the elm (Fig. 246) it is likewise the ovarian wall. In the carrot (Fig. 247) and the Polygonum (Fig. 248) it is an enclosing accrescent calyx. In Piptoptera (Fig. 249) it is two accrescent lobes of such a calyx. In the Zinnia (Fig. 251) a persistent corolla performs the same office. In the hop (Fig. 252) an accrescent bract is made to serve the office of a sail. The fruit of the Cardiospermum represents a class in which the thin pericarp, instead of being expanded into a wing, is inflated into a balloon-shaped receptacle, subserving a similar purpose. Plumes, consisting of the modified persistent calyx, are seen in the Valerian (Fig. 254), where it is present, though concealed by a circinate praefloration, from the flowering stage, while in the Phyllactis it is not developed until

after fructification begins. A plumose style is seen in Pulsatilla (Fig. 255).

We shall next note the cases, perhaps even more numerous, wherein use is made of passing bodies by providing such appendages as shall serve to attach the fruit to them. Fig. 256 represents the fruit of a Rumex, in which the calyx is divided into hooks for this purpose. Fig. 257 shows another species, in which this method is combined with wind transportation, a combination which is very common among the Umbelliferae. The accrescent calyx teeth (awns) of Verbesina (Fig. 258) are adapted to piercing passing bodies, while at the same time the adnate tube is winged. In Bidens (Fig. 259) similar awns are barbed and adhere very

tightly to anything which they may pierce. In the case of the burdock (Fig. 260) it is an involucre which bears such hooks. Similar hooks are found upon the outer wall of the ovary itself in many cases. Sometimes the style is recurved at the apex, thus forming a terminal hook, while at others (Fig. 261) the apex, after performing the stigmatic function, falls away, but leaves a hooked lower joint to become effective in the fruit. The attachment is not always thus secured by means of distinctively piercing appendages. The surface may be rendered adhesive in other ways seen in the minute structures covering the fruit of Desmodium.

We shall next consider another large class of fruits which depend for their transportation upon the possession of edible pericarps or edible portions of them. Such fruits may be eaten with the contained seed, as in the case of the strawberry or small cherries, in which case transportation is effected during the process of digestion of the pericarp; or, as in the case of the peach and plum, the fruit may be too large for such process, depending for transportation upon carriage by a parent to its young. In still other cases they are of such a nature that they can be carried and stored for winter use. The edible portion is in some cases, as in that of the banana, highly nutritious, while in others it is apparently eaten merely for its palatability or for its thirst-quenching properties. Some special form of protection is commonly required for the seeds of edible fruits. That of the peach is enclosed in a hard stone, so that it shall not be abraded as the pulp is pecked or bitten away. Those of the cherry and strawberry are enclosed in similar hard coats, which resist the digestive process as well. The more or less laxative or purgative properties of many fruits doubtless contribute to such protection by the more prompt dejection of the seeds which is brought about by their action.

The origin of the edible portion is various. In the strawberry (Fig. 262) it is the complete torus, and this only. In the blackberry (Fig. 263) such a torus is combined with a partially fleshy ovarian wall upon each of the ripened pistils. In the rose (Fig. 46) it is a similar receptacle, but hollowed, probably with other ele-

ments combined. In the apple (Fig. 264)
it is a fleshy, thickened disk, together with
the adnate calyx lined by it. In the check-
erberry (Fig. 265) it is the calyx only
which becomes fleshy. In the gooseberry it
is the calyx and the entire ovary, but without any disk, while in many other berry-like
fruits it is the ovary alone. In the plum
and cherry not all of the ovarian wall is
edible, its endocarp becoming a putamen.
In the lemon, the papaw (Fig. 267) and the
pumpkin it is the inner portion which is
edible, while the outer is not. In the
watermelon the placentae comprise almost
the whole of the edible portion, while in
the tamarind it is the middle layer of the
ovary.

In all of the above mentioned cases it
is some one or more of the parts of the
flower which eventually forms the edible
pericarp, but there are numerous cases
in which other parts of the plant contribute to or form the whole of such portion.
In the Cashew (Fig. 268) the ovary (a)
enlarges but little, while the petiole (b)

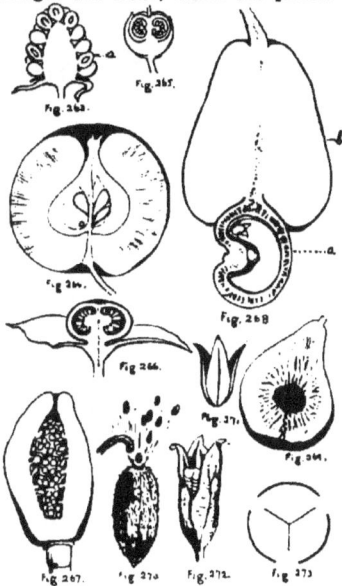

undergoes a great enlargement and becomes edible. In the cactus (Fig. 242)
the end of the branch is hollowed out and
the wall so formed becomes the edible
pericarp of a single flower. In the fig

(Fig. 269) we have a similar hollowed
branch, but instead of being occupied by
a single flower, the wall is lined by an
immense number of them.

Besides the more common methods of
seed distribution referable to the pericarp, which are thus subject to classification, we find numerous special devices
which cannot here be enumerated in detail. Fruits which grow beside or in the
vicinity of streams or other bodies of
water are commonly adapted in some way
for using the latter as a vehicle for transportation. They are frequently of a
rounded form and of considerable weight,
so that upon falling they will roll into
the water, where they are then enabled
to float by virtue of low specific gravity,
due often to the presence in them of large
cavities. The pericarp is in such cases
usually furnished with some means of
protection against the action of the water.
Apparently the thick and woody pericarp
of the Dipteryx is so constructed in order
to avail itself of this method of transportation. The fruit of a species of
Avena is so constructed that by the
change of form and position of its long
awns in dry and wet weather respectively
it is enabled to travel.

Finally we must note that some fruits
are protected by special devices against
transportation. Thus the mangrove possesses a seed which germinates while still
attached to its parent and which does not
sever its connection therewith until the
young plant has descended many feet and
fixed itself into the mud below. The peanut, after anthesis, drives its ovary beneath the surface of the soil, where its
fruit is developed. Plants possessing this
habit are always highly gregarious and in
this way occupy the ground to the exclusion of all other species, thus securing
their perpetuation even though they are
not specially disseminated. The high degree of adaptation secured by the peanut
is further illustrated by its apparent
power to support itself by means of these
buried branches should the parent stem in
any way become severed; a very important protection in view of the highly nutritious character of the herbage which
renders it liable to destruction by grazing
animals.

Mo. Bot. Garden,
1896.

The fixation of many fruits with their contained seeds is secured by a series of devices no less interesting than those which affect their distribution. Fruits like those represented in Figs. 61, 62 &c., are commonly more or less sharpened or narrowed at the lower end, which is much the heavier, so that they shall the more readily penetrate a favorable surface. Their bodies, moreover, are commonly toothed or hispid upward, so that the tendency is for them to sink more and more deeply until properly interred. The fruit of Viscum, whose seed can develop only upon the bark of trees, is intensely adhesive, so that upon falling it can never bound away, but becomes adherent to the first solid body which it encounters.

As a rule, fruits which are provided with special devices for their transportation are not designed for the discharge of the contained seed, which escapes accidentally or develops while still enclosed. Provisions for the discharge of seeds, therefore, ordinarily apply only to such fruits as complete their function at the place of origin. For provisions for the distribution of such plants we should naturally look to the seeds themselves; yet to this rule there are numerous exceptions, for many fruits which never leave the place of growth yet possess various devices for distributing their seeds over a greater or less area by virtue of forces inherent in their pericarps. The common name of the Impatiens, "touch-me-not," is derived from the habit of its fruit of exploding with considerable force, discharging its seeds meantime to a considerable distance. The fruit of Hura similarly explodes, but with such violence as to cause a report like the discharge of a firearm. Elaterium (Fig. 270), during the ripening process, collects by osmosis within its cavity an amount of liquid which exerts a powerful outward pressure upon the pericarp. When fully ripe the slightest contact with another body causes the pericarp to leap away from its attachment, with the production of a hole at its base through which the seeds are expelled with much force.

The ordinary method of providing for seed discharge is by means of a splitting of the pericarp known as Dehiscence. A fruit so splitting is said to Dehisce and is known as a Dehiscent or Dehiscing fruit. Other fruits are called Indehiscent. True dehiscence is longitudinal, although the term is not altogether denied to other forms, provided the line of separation is regular and constant (Figs. 279-282). The parts into which pericarps dehisce are called Valves. The valves may separate entirely or remain attached in various ways. Dehiscence may occur at the ventral or at the dorsal suture or at both. If at the ventral, then the carpel (Fig. 300), or each carpel if it be part of a polycarpellary pistil (Fig. 271), will be

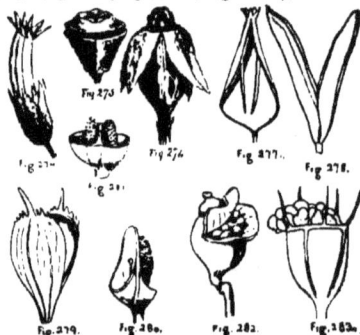

Fig. 273. Fig. 274. Fig. 277. Fig. 278.

Fig. 279. Fig. 280. Fig. 282. Fig. 283.

left entire. If, as sometimes, the polycarpellary pistil have several cells, ventral dehiscence must involve the separation of the carpels by the splitting of their walls or septa, whereas in the one-celled form septa do not exist or are incomplete. Nevertheless the principle is identical in the two cases, and this form is called Septicidal Dehiscence (Fig. 271). In the former of these two cases the carpels, after separating through their septa, are not necessarily open, and unless the dehiscence shall follow the wall into and through the ventral suture, which it rarely does, the dehiscence will be Imperfect and the carpels will act as separate indehiscent fruits. If dehiscence occur at the dorsal suture (Figs. 272 and so on) it must separate the wall of the cell into two parts, and this form is called Loculicidal Dehiscence. By an intermediate form the dehiscence takes place at the point where the septum joins the outer wall (Fig. 273), the Marginicidal. Vari-

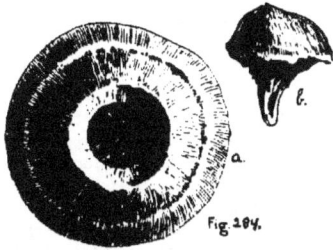

Fig. 204.

ous other modifications and combinations of the two forms may be discovered, but do not call for a notice in this work.

Dehiscence is secured by a peculiar adaptation of the fibres to the other tissues and to the form of the fruit. Various forms of imperfect or incomplete dehiscence are those in which it commences at the apex and fails to extend itself to the base, as in Cerastium (Fig. 274) and Eucalyptus (Fig. 275), or in which it commences at the base and extends only partially toward the apex, as in Jussiaea (Fig. 276) and in Cinchona (Fig. 277). Important pharmaceutical decisions have rested upon the question of basal or apical dehiscence. The true Cinchona barks have all proceeded from species whose fruits dehisce as represented in Fig. 277, while those of the trees yielding the false barks dehisce as represented in Fig. 278.

The manner in which true dehiscence passes into false or transverse dehiscence, called Circumscissile, is well displayed in Figs. 279 to 281, all illustrations of closely related plants. A very curious form of special dehiscence is that of Jeffersonia (Fig. 282).

Dehiscence is not the only method by which fruits open to discharge their seeds. Rupturing fruits are those which open by an irregular line. Some portion of a pericarp may decay quickly, leaving an opening, or the same result may be secured by excessive shrinkage in drying of the pericarp, as in Fig. 283. Openings of the very delicate tissue of some part of this kind are called Pores. Our consideration of this subject will close with an illustration of the fruit of the Brazil-nut (Lecythis, sp. Fig. 284). The apex of this enormously thickened and strongly hardened involucre consists of a small circular portion connected with the remainder by a circle of tissue which quickly decays and becomes movable, thus leaving an apical pore.

CLASSIFICATION OF FRUITS.

A perfect or even fairly satisfactory classification of fruits has never been presented, and it is impossible, except through a complete revision and uniform agreement of terminology, based upon a uniform set of principles. A classification of some sort is, however, an essential in pharmaceutical botany, and such a one is here presented as appears most serviceable to those for whom it is intended.

Among all the various systems which have been proposed two fundamental principles have been observed—first, the morphological structure; second, the physiological features. By the first, fruits have been classed according to the character and number of the parts entering into their formation and the modifications which these have undergone in fructification. By the second, according to the structural characters as seen in the complete fruit, without regard to their mode of origin. As characters of the latter kind exist for the sake of the offices which they are to fulfil it is clear that physiology forms the basis of the latter method of classification. Although it is impracticable to follow either system without some regard to the other, it may be said that to follow in the main the morphological plan is the more scientific, the other the more convenient and the more practical, especially in economic work. This is therefore the plan which is here adopted. Fruits possessing pericarps fitted for transportation will then form the first of our two classes, while those fitted for discharging their seeds in situ upon maturity will form the second.

For a few fruits not readily introduced to this key, and for some exceptions, see the explanations which follow:

a {
  Fruits with pericarp designed for transportation (a).
  Fruits with pericarp not designed for transportation (e).
}

a {
  With fleshy pericarp (Carnose) (b).
  With non-fleshy pericarp (Siccose) (c).
}

b {
  With seeds imbedded in a soft endocarp (g).
  With seeds enclosed in a putamen (h).
}

c { With an enclosing involucre, at least before maturity (l).
     Without an enclosing involucre (d).

d { Vertically divisible into one-seeded parts (i).
     A one-seeded part resulting from such division (j).
     Transversely divisible into one-seeded joints (n).
     Not divisible into one-seeded parts nor the product of such division (k). *

e { Not transversely dehiscent (f).
     Transversely dehiscent (q).

f { Monocarpellary (m).
     Dicarpellary, the valves separating from the placentae (o).
     Not monocarpellary nor dicarpellary, with valves separating from placentae (p).

g { Soft throughout............Berry
     With a soft, tough rind. Hesperidium
     With a hardened rind........Pepo

h { Putamen of bony hardness, Solitary...................Drupe
     Putamen of bony hardness, one of several which are coherent..................Pyrena
     Putamen of bony hardness, one of many which are non-coherent................Drupelet
     Putamen thin and tough.....Pome

i { ....Schizocarp
     (If dicarpellary, with a carpophore ....Cremocarp)

j { Part of a cremocarp......Mericarp
     Not part of a cremocarp.....
         Coccus, Nucula or Nutlet

k { Dehiscent, the valves separating from the two placentae.. Most Silicles
     With thin winged pericarp..Samara
     With inflated pericarp.......Utricle
     Pericarp, thickish in view of its size, not inflated, sometimes winged...................Akene

l { A one-seeded fruit from a glans..Nut
     A non-glumaceous involucre, with contents............Glans
     A glumaceous involucre, with contents. A few.......Spikelets
     A one-seeded fruit from a spikelet....................Caryopsis

m { Dehiscing by one suture only.Follicle
     Dehiscing by both ventral and dorsal sutures..........Legume
     (When spirally coiled.....Cochlea)

n   ....Loment

* Exceptions occur.

o { Elongated.................Silique
     Silicle ....Some Silicles

p   ....Capsule

q   ....Pyxis

The fact, as stated above, that custom has not been uniform in the application of the principles of classification leading to the above terms, so that the latter are not employed in the same sense in botanical writings renders it necessary that such a key as that presented should be supplemented by a detailed consideration of the limitations and modifications of each class of fruits.

The Berry (Figs. 242 and 266).—A fruit with a pericarp fleshy throughout, with the exception of the epicarp. Good illustrations are the grape and the belladonna. In these the fruit contains no cavity and the seeds are imbedded in a soft pulp. This is the typical form, from which we see a variation in the tomato, in the direction of a central cavity which in the capsicum becomes complete. The latter is frequently called a capsule and connects the berries with the latter class, but it is more properly grouped with the berries. A similar modification, though more slight, is found in the gaultheria and the cranberry. The term has also been applied to the pomegranate and similar fruits, but these, however soft within, possess a distinctly hardened exocarp and are not true berries. As will be seen farther on, comparatively few of the fruits which are designated as berries in common parlance are really such. The berry may be one or more celled.

The Hesperidium (Fig. 285).—A berrylike fruit with a soft, but tough rind. The term has never been applied to other fruits than those related to the orange and lemon. They are several celled.

The Pepo (Fig. 286).—A berrylike fruit in structure, usually hollow and with an indurated rind. It is one celled. Good illustrations are the pumpkin and melon, and the application of the term is by most authors restricted to the fruits of that family (the Cucurbitaceae); but it is entirely proper to extend it to such entirely similar fruits of other families as the Calabash (in the Bignoniaceae), and the Papaw (in the Papayaceae).

The Drupe or Stone Fruit (Fig. 287).—A

fruit with a sarcocarp and epicarp and a single thick bony putamen. Although typically one celled and one seeded, the term is applicable to similar fruits with

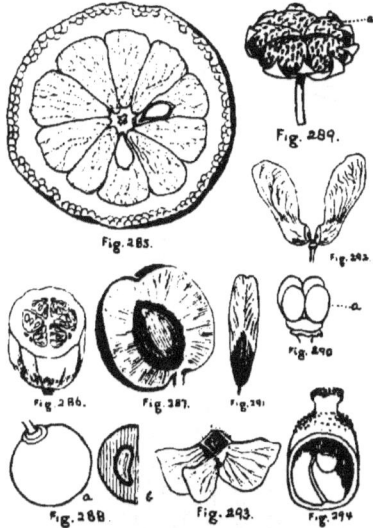

Fig. 289.

Fig. 285.

Fig. 286. Fig. 287. Fig. 291.

Fig. 288. Fig. 293. Fig. 294.

several cells enclosed in a single sarcocarp, but each seed possessing its own putamen. Each putamen with its own seed is then called a Pyrena or Pyrene. Familiar illustrations of the typical drupe among medicinal plants are the prune, sumach and pepper, and of the several celled form that of the Rhamnus. As in most other classes of fruits we find here a gradation into other classes, most commonly into the Schizocarp. A peculiar fruit, in its general structure related to the drupe, is the so-called legume of the tamarind, which possesses an exocarp similar to that of a pepo, a distinct edible sarcocarp and a crustaceous endocarp or putamen containing several seeds.

Pyrena (Fig. 288).—(Already considered under drupe).

The Drupelet (Fig. 263, a).—Differing from the Pyrena in that it possesses not only its own separate putamen, but sarcocarp as well. It is one of many small drupes belonging to an aggregate or multiple fruit.

The Pome (Fig. 264).—A fleshy fruit with a thin chartaceous or cartilaginous putamen. It is several celled. The term

is commonly restricted to fruits related to the apple.

The Schizocarp (Figs. 247, 289 and 290).—The typical schizocarp should be defined as a fruit which divides septicidally at maturity into one seeded carpels. Because, however, schizocarps frequently vary in the constancy and completeness with which they undergo this process, they are defined as "divisible," rather than "dividing." There are, moreover, cases in which they divide into one seeded parts of carpels. The comprehensive definition, therefore, should be fruits septicidally divisible at maturity into one seeded parts. Schizocarps are commonly provided with appendages for wind transportation or for transportation by mechanical adhesion to passing bodies. Those forms which, as above stated, are intermediate toward drupes are to be classed in one or the other class, according to whether such appendages for distribution or that of an edible pericarp is the more pronounced. Even schizocarps which are not cremocarps may possess a carpophore, as in the geranium, though commonly they do not.

The Cremocarp (Fig. 247).—A di-carpellary schizocarp, the carpels attached toward their summits to a slender carpophore, from which they usually only incompletely separate at maturity. The term is restricted to the fruits of the Umbelliferae. They are commonly provided with appendages for fixation to passing bodies, frequently for wind transportation, and not rarely combine these two methods of distribution. (Conium, Celery, &c.). There is no class of fruits which possesses a greater importance in pharmacy, and hardly any whose histological features are of greater interest. The plane of separation is called the Commissure, a term applicable to a similar plane in other fruits. (See Mericarp.)

The Coccus Nucula or Nutlet (Figs. 289 and 290, a) is one of the divisions of a schizocarp, and its nature has been explained in considering that group. The term nutlet is commonly applied when the pericarp is hard and close to the seeds.

The Mericarp (Fig. 247, a).—One of the halves into which a cremocarp is divisible. Occasionally they are self-separating at maturity, but usually only incompletely so. They are one seeded and possess a completely adnate calyx and disk. The pericarp almost uniformly possesses external appendages in the form of five or nine ribs. When nine, they are commonly of

The Samara.—An indehiscent fruit with a winged pericarp. They are commonly one seeded, as well as one carpelled, but may be more. Typically, it is the ovarian wall or the tube of an adnate calyx which develops the wing, but there is no reason

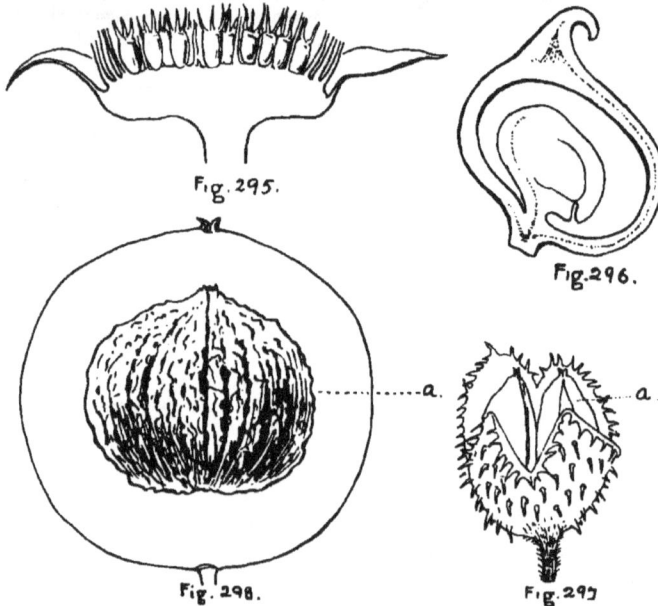

Fig. 295.

Fig. 296.

Fig. 298.

a.

a

Fig. 297

two forms, alternating with one another. A part or all of them are much subject to extension into variously appendaged wings. Internally the mesocarp is almost uniformly traversed upon both the faces and the backs of the carpels by tubes called Vittae, commonly with suberous walls and filled with volatile oil. The dorsal vittae alternate in position with the ribs. Upon thin transverse sections these oil ducts or vittae appear as perforations, and as to their number and position serve the most important purposes in diagnosis and identification, as do also the ribs. Mericarps are of three classes:—(1) The Coelospermous, characterized by the possession of a concave face; (2), the Campylospermous, characterized by the possession of a longitudinally grooved face, and (3), the Orthospermous, possessing a plane face

The Silicle.—(See Silique.)

why the term should not be extended to include similarly transportable fruits with wings consisting of the accrescent limb of a calyx (Figs. 247 and 248), or corolla (Fig. 251), or a surrounding alate bract (Fig. 252).Commonly the samara possesses but a single wing, unilateral, as in the ash (Fig. 291), or circular, as in the elm (Fig. 246), but not rarely more than one wing is present, as in the maple (Fig. 292), or many Malpighiaceae (Fig. 293).

The Utricle (Fig. 294).—A one seeded indehiscent fruit, the seed enclosed in a thin, bladdery or inflated pericarp. It is commonly one celled, but occasionally several celled. Ordinarily they eventually become irregularly ruptured, but in a few forms there is a regular ventral opening, approaching toward dehiscence.

The Akene or Achenium (Figs. 61 to 67 and 296).—A small, indehiscent, one seeded, seedlike fruit, the pericarp somewhat thickened and entirely distinct from the enclosed seed. The akene varies in many directions toward other fruits. In many cases the pericarp is inclined to be fleshy and in a few it tends toward dehiscence, thus simulating a follicle. Some forms of the akene are distinctly winged, so that they might but for the relationship of the species yielding them to akene producing species be with equal propriety classed as samaras. They are in nearly all cases provided with some means for securing wind transportation or for attaching themselves to passing bodies, and yet there are numerous cases in which all such appendages have become entirely obsolete. For these reasons it becomes a matter of extreme difficulty to frame a definition at once comprehensive and delimiting for this group. The inferior akene is sometimes distinguished by the term Cypsela (Figs. 61 to 67).

Note should here be taken of the fact that the latter is characteristic of that largest of all families, the Compositae, in which the akenes of the head are massed and partially, or sometimes completely, surrounded and enclosed by an involucre, the whole constituting a multiple fruit to which the name Hypanthodium (Fig. 295) has been applied. The Hypanthodium varies greatly in its characters. Although usually multiflorate, it is commonly few, or even in rare cases one flowered. In those cases in which the involucre completely encloses the achenia it is commonly appendaged for distribution in an entire condition, as in the burdock. This condition connects the Hypanthodium with the glans and the contained achenia with the nut. Indeed, it is almost impossible to distinguish structurally between the fruits of the Xanthium and that of the Fagus.

The Glans (Fig. 297 and 298).—A fruit consisting of an accrescent and partially or (commonly completely) enclosing involucre containing one or more nuts. The involucre may be dehiscent as in the chestnut and hickory nut, or indehiscent, as in the Brazil nut and black walnut. In some of its forms, moreover, the involucre of the glans tends to become fleshy. Inasmuch, however, as the design of such pseudo-fleshy pericarps is not that of subserving transportation by their food prop-

erties, they are more appropriately regarded as non-fleshy. While depending like the grasses upon their gregarious habits for perpetuation, nut-yielding plants apparently in many cases are distributed by the rounded form of their coats and the readiness with which they are transported by flowing water.

The Nuca or Nut (Figs. 297 and 298 a).— The relationship of the nut and its glans to the akene and its hypanthodium has already been pointed out. The nut is in all cases much larger than the akene and its pericarp commonly much thickened and very hard.

The Spikelet (Fig. 299).—A fruit possessing a glumaceous involucre and pertaining to the Gramineae (grass family) and related orders. This class of fruits, like the glans and nut, connects those fruits which are adapted to transportation with those which are not. Although in general these plants depend for their perpetuation upon a highly gregarious habit rather than upon provisions for distribution of their fruits, yet the spikelets of some grasses are unmistakably so designed, and are transported with the caryopsis enclosed in the glumes.

The Caryopsis or Grain (Fig. 299 A).—A seedlike fruit produced in a spikelet, the ovarian wall and the seed coats closely adnate.

The Follicle (Fig. 300).—A monocarpellary fruit dehiscing by one suture only, except in rare cases the ventral.

The Legume (Fig. 301).—A monocarpellary fruit, non-fleshy and dehiscing by both ventral and dorsal sutures. Notwithstanding the definition thus given, we have to record the fact that in accordance with a different principle and construction, the title includes all fruits of the natural order Leguminosae. It therefore becomes necessary to point out that the fruits of this family are extremely variable, and this in directions which frequently carry them widely away from the structural characters of the legume. The peculiarities of the tamarind have already been pointed out. In the fruit of the Inga the dehiscent legume is filled with a large amount of juicy pulp, in which the seeds are embedded. In other species this pulp is replaced by one of a powdery consis-

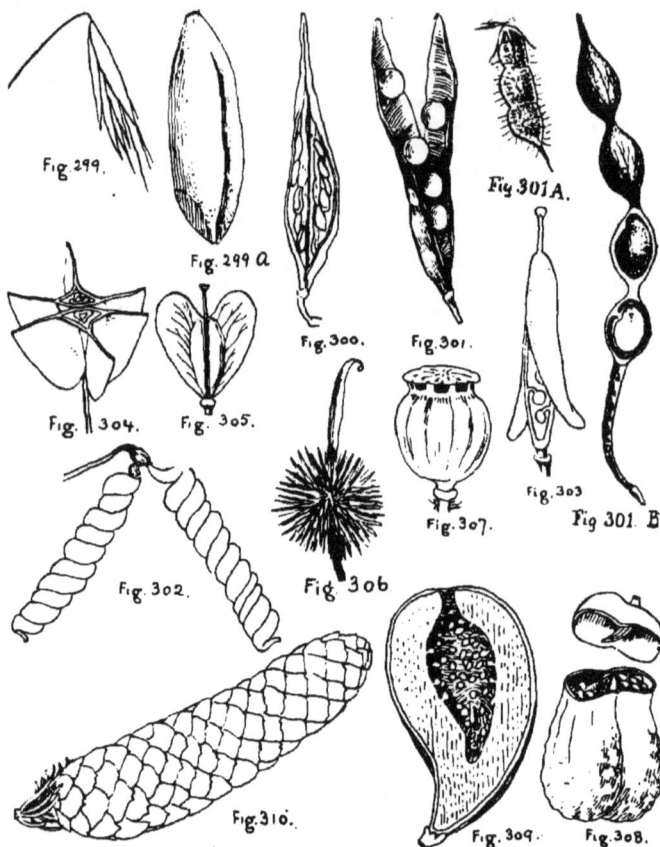

Fig. 299.

Fig. 299 a.

Fig. 301 A.

Fig. 300.

Fig. 301.

Fig. 304.

Fig. 305.

Fig. 303.

Fig. 307.

Fig 301. B

Fig. 302.

Fig. 306.

Fig. 310.

Fig. 309.   Fig. 308.

tency, while in others it is fleshy or sub-corneous. A great many legumes of his family are not only indehiscent, but winged and one seeded, and thus are true samaras. The fruit of the Dipteryx is one seeded and tardily dehiscent, but the pericarp is enormously thickened and woody or corky. That of the Cassia Fistula has its seeds enclosed in a pulp and partly separated from one another by transverse septa. It is thus apparent that many legumes pertain to our first, rather than to our second, division. Two distinctive forms of the legume have become dignified by the application of special names, as follows:—

The Loment (Fig. 301 A, B) is a legumin-ous fruit which may or may not be dehiscent, but which is separable at maturity by transverse divisions into one-seeded parts. In the Desmodium these parts are adapted to fixation to passing bodies, or occasionally much flattened and expanded to act as samaras. In the sophora the joints are smooth, hard and rounded, and highly elastic, so that in falling upon the stony soil they are adapted to bounding and rolling to a considerable distance. The term loment has also been extended to include those siliques which display a similar character.

The Cochlea (Fig. 302) is a legume which is spirally coiled.

The Silique (Fig. 303) is a di-carpellary

dehiscent fruit, the two valves separating from the margins of the placentae at maturity, leaving the latter attached to the torus and to a false septum, which divides the silique into two parts. The principal modification of the silique proper is into the lomentlike form which we have already considered. This class of loment-producing plants are commonly found in the vicinity of water, and their fruits are adapted to transportation by this method. A more important modification is into the

Silicle (Figs. 304 to 306), which differs from the silique not only in being short and broad, but in possessing ordinarily some form of adaptation to wind or other transportation, thus belonging in our first class.

The Capsule (Figs. 240, also 271 to 283).—The typical capsule is to be defined as a di to polycarpellary longitudinally dehiscent fruit. From the typical form, however, it varies in several directions to such a degree as to render it impossible to frame a perfect definition. The capsule of the poppy (Fig. 307) opens by a number of small pores at the summit and this is true of many other forms. In other cases the mode of opening is by various forms of irregular dehiscence intermediate between the longitudinal and the circumscissile. Finally, we must note that many fruits, like those of some species of Passiflora, which possess no regular or natural method of opening are still classed as capsules by systematic botanists.

The Pyxis (Fig. 308).—A circumscissilly dehiscent fruit.

The Syconium.—A fruit consisting of a hollow branch, becoming fleshy, its inner surface the receptacle for many small, one-seeded akene-like fruits.

The Aeterio (Figs. 262 and 263).—An aggregate fruit, with an accrescent fleshy torus and many crowded pistils.

The Strobile (Figs. 310 and 311).—A multiple dry fruit, its elements in the form of imbricated scales.

The Galbalus (Fig. 312).—A fruit similar to the last, but the scales fleshy or much thickened above, so that the form becomes more or less globular.

In conclusion it may be remarked that to assign a name to a fruit is insufficient in most cases, especially those of aggregate and multiple fruits, to designate its character. The title must be supplemented by more or less of a description.

CHANGES IN THE OVULE.

As in the case of the parts entering into the formation of the pericarp, so in that of the part forming the seed—namely, the ovule—it is well to precede our study of the changes which it undergoes by consideration of the objects to be attained thereby. The essential feature of the seed is the possession as one of its parts of a more or less rudimentary plant, developed from the fertilized oosphere, and known as the Embryo. During the period intervening between the beginning and the completion of seed-formation the embryo requires nourishing, and provision for this constitutes the first requirement of the process. The further development and growth of the embryo between the time of germination and that of absorption by it from the external world calls for additional nourishment. This can be met only by the storage as a part of the seed of an additional food supply. Protection of the seed contents during its development is only partially afforded by the pericarp, and this office is supplemented by the coverings of the seed itself, while its self-protection between the periods of maturity and germination is a manifest necessity. The transfer of the mature seed to the point of germination, or its Dissemination, and its fixation in a favorable site have already been referred to. We have seen that in many cases these offices are not provided for by the pericarp, and we must look for such provision to the seed itself. The parts of the seed by which these several offices are performed we find to be as follows:—The food supply during the germination of the macrospore and development of the gametophyte we have seen to be the portion of the nucellus external to the embryo-sac. Usually more or less of this material remains during at least the earlier period of the development of the embryo and contributes to the nourishment of the latter. Occasionally it persists even in the seed condition. It is known as the Perisperm. Inside the embryo-sac a further store of nutriment is caused to develop as a result of fertilization, this constituting the chief supply of the growing embryo. More or less of this also may persist, and usually does, upon the maturity of the seed. It is known as the Endosperm. As the embryo develops it stores within its own body more or less nutriment. At maturity we may find that the entire store of nutriment has thus been transferred to the body of the embryo, and the seed is said to be Exalbuminous, or we may find more or less endosperm or perisperm, or both, when the seed is said to be Albuminous, and this nourishment is known as the Albumen. In only a few seeds used in medicine does the albumen consist in any part of perisperm. The chemical nature of the albumen is extremely variable. Protection to the embryo may be afforded by the albumen when that is of the required consistency or composition, the conditions of the latter being a mere parallel of those already considered under the subject of the pericarp. More frequently, however, it is secured entirely through the coats of the seed. These may correspond to the coats of the ovule, though usually the primine is found to have disappeared. When it persists it takes the name of Tegmen, or Endopleura, the secundine becoming the Testa, or Exopleura. Rarely the secundine also disappears and the seed is Naked. The seed will also be naked when produced from a naked ovule. When one or both of the coats persists, the point where the foramen existed becomes the Micropyle. Frequently the development of a new coat external to the others is induced by fertilization, and this is known by the general name of Aril. If it develops from the chalaza or a lower point it is called an Arillus, or True Aril; if from the micropyle, an Arillode, or False Aril.

The provisions of the seed for securing dissemination are in most respects comparable with those affecting the pericarp. Wind distribution is pre-eminent; that by fixation to passing bodies is frequent, and that by means of an edible coat is rare.

The fixation of seeds disseminated without the pericarp is favored by their small size, enabling them readily to enter crevices and cavities, and by peculiarities of surface which favor the same process.

The large number and importance of medicinal seeds lend great importance to their study by the pharmacognosist, and this is especially true of the histology of all their parts. Inasmuch, however, as the subject of histology has been referred to a separate portion of the work, we shall here consider only such characters as can be distinguished by means of an ordinary lens.

The hilum is in most cases readily perceptible, but is occasionally found only by minute examination. It is to be studied as to its position, size, form, surface and color. Its position is sometimes fixed with reference to the form of the seed, as at the larger or smaller end, upon the broad side or on the edge. as well as with reference to the micropyle, adjoining it, at the opposite end or at some intermediate point. It is the last mentioned character which determines the class of seed as to its tropism. In size the hilum may be a slight point or covering a considerable portion of the surface. Its form is frequently characteristic as heart-shaped (Fig. 314) or linear and channeled as in Fig. 315. Its color frequently differs markedly from that of the remainder of the seed.

The rhaphe, extending from the hilum to the chalaza when these do not coincide, is ordinarily not perceptible upon the surface. When it is so, as in Figs. 316 and 316A, its appearance is of great diagnostic value and must be closely scrutinized. The chalaza in its simple form calls for no especial attention. If, however, an enlargement appears at this point (the Strophiole, Fig. 317), it must not be overlooked. The strophiole may develop into the arillus (Fig. 317A), a partial or complete covering, and its characters call for the same attention which is requisite for the testa.

The testa is not wanting in any medicinal seeds. In general it is not closely adherent to the underlying tissue, and it can be readily removed. In its thickness, consistency, surface, color and appendages it yields important pharmacognostical characters. It may be pitted (Fig. 318), tuberculate (Fig. 319), reticulate (Fig. 320) or hairy (Fig. 316), and the minute characters of its pits, tubercles, ridges or hairs must not be overlooked. It may be dull or shiny, and its color may

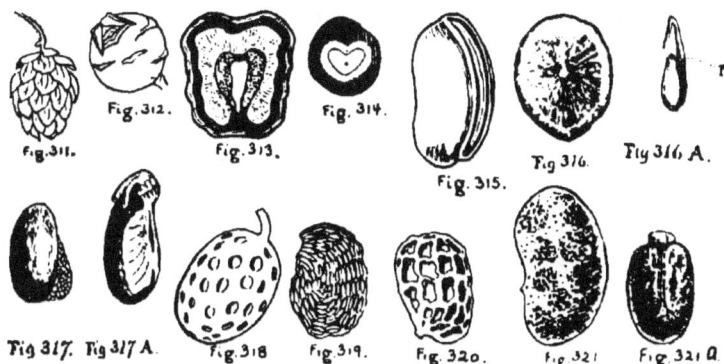

Fig. 311. Fig. 312. Fig. 313. Fig. 314. Fig. 315. Fig. 316. Fig. 316 A.

Fig. 317. Fig. 317 A. Fig. 318. Fig. 319. Fig. 320. Fig. 321. Fig. 321 A.

be uniform or variegated (Fig. 321). Its lustre or shade of color is frequently of the greatest assistance in determining the age, freshness, mode of preparing or preserving or other conditions on which the comparative medicinal quality of the seed depends. The enlargement at the micropyle (the Caruncle, Fig. 321A) calls for the same scrutiny as the strophiole. It may be variously appended (Fig. 323), and like the latter it may extend into a partial or complete covering, the arillode. The arillus, or arillode, may be partial, as in Fig. 322, or complete, as in the seed of the Euonymus.

Appendages to the seed do not always take the form of an aril of either class, nor is their origin confined to the points from which the aril develops. Either as aril or appendage from the general surface it exhibits a great variety of form, of equal importance with those which characterize the pericarp. Forms of especially frequent occurrence are exhibited by Figs. 324 to 329. Important distinctions sometimes exist between seeds bearing similar appendages, as regards the points from which the latter originate, as in the case of strophanthus, false and true.

The general form of the testa is, of course, that of the seed, and calls for terms applicable to the forms of solid bodies (Figs. 330 to 337).

The tegmen when present is extremely thin and tightly adherent to the nucellus, following closely all inequalities upon the surface of the latter, and occasionally having its intruded folds caught between the forming masses of the albumen and

discernible upon section of the latter as slender veins, giving us the so-called Ruminated Albumen (Fig. 338).

The albumen is characterized chiefly by its consistency, being bony, as in the ivory nut and date; horny, as in nux vomica; oily, as in the castor oil and cacao; fleshy, mealy, etc. In sectioning the seed note should be taken of the presence, number, position and forms of any cavities which may exist in them.

The embryo calls for the most thorough and minute study as a basis for systematic work, though for the pharmacognosist only the more important details of its general structure need be considered. It has already been stated that it consists of one or more phytomers. The part first formed is the internode, which becomes the Caulicle (Figs. 339 to 340A, ca.), in old works denominated the "Radicle." The Radicle (in the last named figures) is the extreme tip of the caulicle, which points always in the direction of the micropyle. From this point the root is to be developed. The embryo may consist of nothing further than the caulicle, and even this may be of the most elementary character. Ordinarily, however, there develops at the node (the point opposite to the radicle) one or more Cotyledons, or Seed-leaves (cot, in the figures). Most seeds which possess but a single cotyledon (Fig. 341) are grouped together in a division of the Angiosperms, which for this reason are called Monocotyledons, those with two in the Dicotyledons. A few plants, mostly Gymnosperms, are polycotyledonous (Fig. 342). The highest plants of their respective groups develop a second phytomer lying between the coty-

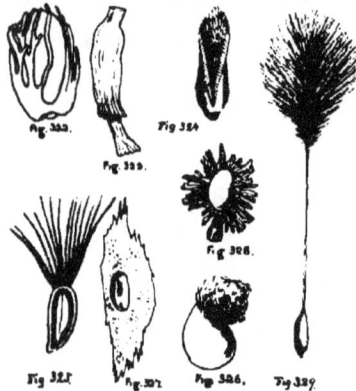

Fig. 322.   Fig. 324.

Fig. 323.

Fig. 326.

Fig. 321.   Fig. 327.   Fig. 326.   Fig. 329.

ledons, or if there be but one cotyledon mostly enwrapped by it. This is the Plumule (Fig. 343 pl.) which shows the same variation in the degree of its development as that which characterizes the lower. When its leaves are developed they bear a closer resemblance, as in the figure, to the mature leaves of the plant than do the cotyledons, following out the law referred to in our introduction. The plumule commonly pertains to exalbuminous seeds among the dicotyledons.

Terms used to indicate different directions of the radicle refer to its direction with relation to the fruit, its direction in relation to the micropyle being, as has

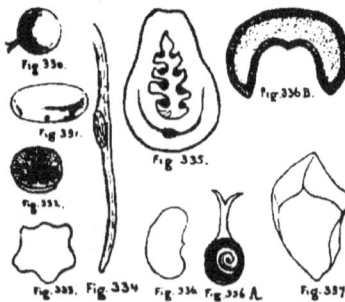

Fig. 330.

Fig. 336 B.

Fig. 331.

Fig. 335.

Fig. 332.

Fig. 333. Fig. 334   Fig. 334   Fig. 336 A.   Fig. 337.

been stated, always the same. It is Ascending when it points toward the apex of the fruit, Descending when in the opposite direction, and Horizontal when intermediate. The latter form is Centrifugal when pointing toward the periphery,

Centripetal when toward the axis.

The position of the embryo with reference to the albumen is always highly characteristic. It is Axile or Centric when in the centre of the albumen (Figs. 340 and 340A), whether straight or curved; Eccentric when within the albumen, but outside of its centre (Fig. 344); Peripheral when lying upon the surface of the albumen. In the latter position it may be straight or simply (Fig. 345) or circinately (Figs. 346 and 346A) coiled. The relative sizes of the embryo and the albumen vary from those in which the former is a mere speck in a large mass of the latter to that in which the proportions are reversed, or in which the albumen is entirely wanting. The embryo should in all cases be dis-

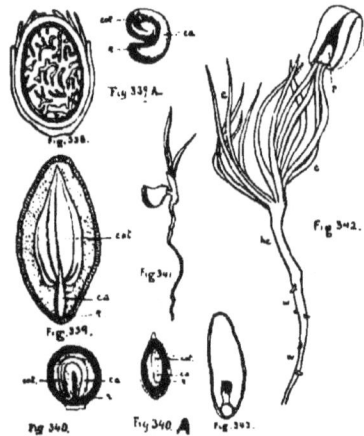

Fig. 338 A.

Fig. 342.

Fig. 338.

Fig. 341.

Fig. 339.

Fig. 340.   Fig. 340 A   Fig. 343.

sected from the contiguous parts and the relations of its parts to one another made out. It may be straight, variously curved, crumpled (Fig. 347) or variously folded. In the latter condition the radicle may be brought into juxtaposition with the edges of the cotyledons (Accumbent, Fig. 348) or with the face of one of them (Incumbent, Fig. 349). One cotyledon may enwrap the other (Fig. 350). When a single cotyledon partly encloses the greater portion of the remainder of the embryo it is sometimes called the Scutellum (Fig. 351). Some of the terms applicable to the consistency of the albumen are also applicable to that of the cotyledons.

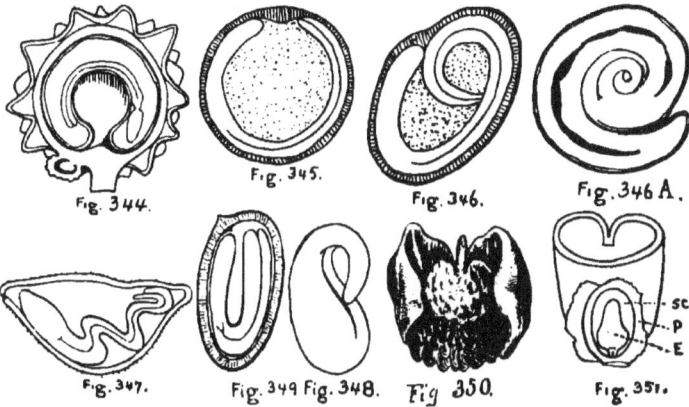

Fig. 344. Fig. 345. Fig. 346. Fig. 346 A.
Fig. 347. Fig. 349 Fig. 348. Fig. 350. Fig. 351.

Finally, the pharmacognosist will find it of importance in the case of seeds possessing a characteristic taste to inform himself as to the part, if any, to which such taste is restricted.

With the production of the seed, containing a distinct living individual separated from the parent and fitted for independent existence, reproduction can strictly be considered as completed, although the progeny is still in its infancy and its form not yet perfect. Its analogy to the bud is apparent. Each consists of one or more vegetative units ready to develop under proper conditions into a perfect semblance of the parent, and each is provided with a store of prepared nourishment to sustain it until able to manufacture such for itself. The distinction is in the radically different modes of origin and in the structure, leading to different powers of reproduction.

GENERAL STRUCTURE OF ROOT AND STEM.

The development of the embryo commences with the division of the fertilized oosphere into two cells, each of which grows and becomes capable of itself dividing similarly. The result of such cell propagation is the production of a tissue and of a body which becomes elongated through successive transverse divisions of its cells, or certain of them, and broadened by their longitudinal division. So long as the cells produced are the same in kind the body consists of but one tissue; but through differentiation and specialization among them different tissues are soon developed. The power of cell-division and growth is lost by most tissue after a time, while in other parts it persists permanently. Any tissue or portion of tissue which possesses such power is called Meristem or Meristematic Tissue. Tissue may cease finally to exert meristematic power, or it may resume such power after a time. All meristematic processes cease upon maturity of the seed, recommencing with germination. The point reached in the development of any plant-body in the embryonic condition—that is, at the maturity of the seed—does not depend in any degree upon the amount or kind of tissue or tissues developed, but altogether upon the habit of the particular plant. In some embryos tissue differentiation cannot be seen to have taken place at the time of separation from the parent, while in others it has progressed very far, though never (unless germination has occurred) to the production of a true root. It is impossible, therefore, to fix upon any particular developmental stage of stem structure as distinguishing the ungerminated embryo from the germinated plantlet. In the following sketch of its development, then, no note is taken of the resting period in the seed state, but the process is followed as though it were continuous from fertilization into the mature condition of the plant. The phenomena of germination are not important from the standpoint of pharmacognosy, and a mere outline of them is here given.

Animation is probably not entirely suspended during the resting period of the seed. That is, there is an apparent interchange of substance, due to vital action, between the seed and the surrounding atmosphere, although extremely slight, so long as the former possesses its vitality. Germination depends upon (1) a specific temperature, varying for seeds of different species and for those of the same species when they have become habituated to essentially different climatic conditions; (2) a specific saturation, also varying with different seeds—that is, the absorption of an amount of water bearing a fixed ratio to the weight of the seed: (3) a partially fixed degree of light exclusion; (4) the presence of free oxygen. Under these conditions ready prepared nutriment is dissolved, other forms become digested by special vegetable ferments (Enzymes) present, heat is developed, cell propagation and cell growth take place and the development and growth of a plant from the embryo commence. By the growth of the embryo the radicle is protruded through the micropyle, the rest of the body soon following. The radicle, if it does not already point directly downward, turns in that direction and develops into a root. The opposite end of the embryo, if it does not already point upward, turns in that direction and develops as the apex of the stem. The stem above the cotyledons is called the Epicotyl, that below them the Hypocotyl.

The cellular nature of development and growth demands a general knowledge of histology for their understanding, so that we shall here consider, so far as possible, only the gross results of the processes, or such characters of the root and stem as can be demonstrated by other than histological methods. Such references to cellular structure as are here necessary are given rather figuratively than technically. The mode of growth in root and stem, and the structures resulting, are sufficiently different to require separate treatment. Although the

forms of structure here considered as applying to the root concern only flowering plants and the very highest of the cryptogams, yet the description is applicable to all roots used in medicine.

STRUCTURE OF THE ROOT.

Upon examining a transverse section of the root in its rudimentary condition (Fig. 352) it is possible to distinguish three bodies of tissue exhibiting characteristic differences in their cellular elements. The central portion is occupied by a solid cylinder called the Plerom (a). Outside of this there is a hollow cylinder called the Periblem (b), and still outside of this and upon the surface of the root a second hollow cylinder, the Dermatogen (c). The last mentioned develops a primary covering called the Epidermis. The epidermis consists in its earliest stage, and therefore at the very tip, of a number of layers of cells which protect the apical growing point of the root, and is therefore called at that point the Root-cap. Toward the summit of the root-cap the outer layers successively wear off or are cast off, so that the epidermis becomes reduced to a single thickness of cells. Here it frequently develops a dense covering of Root-hairs which adhere tenaciously to the soil and perform various processes connected with absorption. For this reason the epidermis of the root is known as the Piliferous Layer. Still further up these hairs have fallen away and the single layer, after slight modifications, becomes converted into the epidermis proper. This has a variable duration in different plants and is consequently found covering the root for a greater or less distance upward. Almost always its duration is very short. It either disappears altogether, being replaced by a structure (Periderm) developed from the periblem, or in rare cases itself develops into the periderm.

The periblem of the root develops into the Cortex, consisting of a number, often a large number, of layers of cells. Its outermost portion, usually of one layer of cells, presents a different appearance from the subjacent layers, and is the Hypodermis (Fig. 354b), in the case of the root becoming the Exodermis. The hypoderm lies against the inner face of the epiderm (a), while that persists, becoming

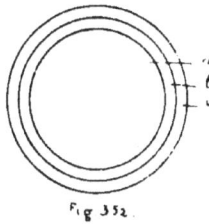

Fig 352.

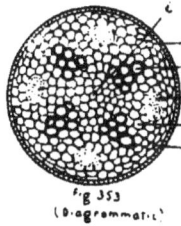

Fig 353.
(Diagrammatic.)

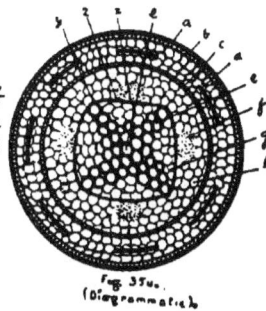

Fig 354.
(Diagrammatic.)

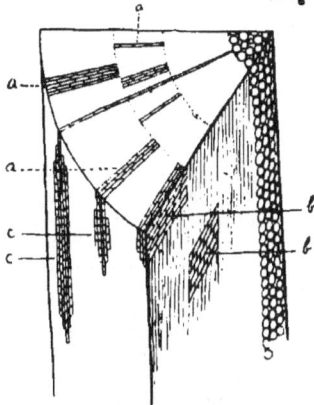

Fig 355. (Diagrammatic.)

afterward the superficial layer, and persists for a longer or shorter period. Its characteristics are of great importance in histological determinations. The inner most layer of the primary cortex is even more distinct in appearance than the hypoderm, and is the Endodermis (c). It lies in contact with the outer surface of the structure developed from the plerom. The production of primary cortex is quickly completed. If then the growth inside of it continues indefinitely it, in most plants, involves the destruction and disappearance of the primary cortex, which must be replaced by some other covering. A new meristematic region must then be established for the purpose of manufacturing such a covering. This almost always arises in some part, and it may be in any part, of the primary cortex. It is the Phellogen. The phellogen may be in the form of a continuous circle

or in that of blades or plates (d), variously placed and directed. Upon its outer surface the phellogen develops corky tissue, the Periderm, and upon its inner a secondary cortex, the Phelloderm. Occasionally it will produce only periderm or only phelloderm. As the periderm becomes impervious to the nourishing fluids it, and all tissue exterior to it, must die, and may be cast off, a new phellogen then appearing further toward the interior to form a new periderm, so that we may have successive periderms—the primary, secondary, and so on. This process is comparatively rare in the case of the root, very common in that of the stem. In such case the corky layers which become successively superficial constitute the Bork or Rhytidoma. Bork is called Ring-Bork when it forms a cylinder, Scale-Bork when it occurs in detached plates. It must be noted that the origin of the bork, and, as will be shown later, its structural nature dependent thereon, will depend upon the depth at which the phellogen develops. The same feature will also determine the amount and character of the tissue, if any, existing between it and the structure developed from the plerom. No tissue developed directly or indirectly from the periblem is in the form of distinct and regular bundles of vessels, though irregular and isolated or anastomosing tubes are frequently developed by it.

The essential characteristic of the body developed from the plerom of the root is that it is invested by the endodermis and is free from any other endodermal development in any part. It therefore constitutes a Stele (all inside of c), which in the root is always in the form of a Central Cylinder. The plerom exhibits at first only slight differences in the appearance of its cells (Fig. 352a), and a transverse section of it viewed with the microscope might be figuratively compared to looking down upon a honeycomb built in a cylindrical tin box, the latter representing the endodermis, and in longitudinal section to a longitudinal section through the same. Further away from the tip, however, it would be found that groups of its cells (Fig. 353 e and f) had elongated in a longitudinal direction and these, to continue our illustration, might be compared to bundles of pencils or quills set in the honeycomb. Mingled among the elongated cells of the bundle, however, are many which have not elongated. These bundles would be arranged in a circle, separated from one another by more or less of the honeycomb tissue, these separating portions corresponding to the Medullary Rays of the Stele (g). From the endodermis they would be separated by one or more continuous circles of the honeycomb cells, corresponding to the Pericycle, or "Pericambium" (h). For a time there would also be left a central portion (i), consisting of unchanged cells. The elongated cells, which constitute the important elements of the bundles, are joined end to end with other similar ones still further up in the older part of the structure. At first the end walls of these abutting cells separate their cavities from one another, but later they disappear in some, becoming perforated in others, so that the cavities become more or less continuous, forming the Vessels, extending throughout the root and into and through the stem above. The bundles thus formed are seen to be of two kinds, alternating in the circle. Each of those of one kind (Fig. 354e) extends gradually toward the centre by the successive conversion of the original cells left there into vessels or into cells associated with the vessels of the bundles. Upon meeting there, they of course cut off the previously existing central communication between the medullary rays, which are now left as isolated plates or wedges between the bundles. These bundles,

which meet at the centre, are known as the Xylem-bundles, or Wood-bundles, constituting the woody portion of the root. In a few plants which we have to consider, the Gymnosperms, no series of cells lose their end walls as above described so as to become converted into continuous tubes, the Ducts, though some of them connect by perforations.

The other bundles (Fig. 354f) which have been described as alternating with the xylem, or wood-bundles, possess as their important element those cells which become connected by perforations in the form of sieves, and are known as the Phloem-Bundles. Collectively they form what is known as the Sieve-tissue, or Cribrose-tissue, of the plant, and their intercommunicating tubes are the Sieve-tubes. This tissue characterizes the Gymnosperms as well as the Angiosperms. The phloem-bundles do not extend toward the centre, as do the xylem-bundles, but stand isolated, each between two medullary rays which separate it from the xylem-bundle upon either side. In connection with the ducts, or their equivalents in the gymnosperms, and the other tissues of the xylem-bundles, develop strong fibres, the wood-fibres, while in connection with the sieve-tubes and other tissue of the phloem-bundles usually develop very similar fibres, the Bast-fibres. The phloem-bundles therefore ordinarily become Bast-bundles. Vascular bundles in which fibres develop are known as Fibro-vascular bundles.

The condition now reached by the root constitutes the completed primary structure of its stele. With the production of the primary structure growth and increase in thickness may cease (Monocotyledons), in which case the periderm changes which we have recorded will not occur. On the other hand, secondary growth may take place, in which case those changes are more or less completely induced. In such case the cells touching the phloem-bundles upon their inner faces and upon their sides become meristematic and proceed to produce xylem-tissue upon their inner faces and secondary phloem upon their outer, in contact with the primary tissue of that kind. Each such arc of meristem (Fig. 354x) becomes the Cambium of that bundle. At the same time the cells lying in contact with the outer surfaces and with the sides of each xylem-bundle similarly become a cambium for that bundle (y), and sometimes produce secondary xylem, upon their inner faces, in contact with the primary xylem there, and secondary phloem upon their outer faces. By these processes each bundle, previously consisting of one kind of tissue, therefore an incomplete bundle, comes to consist of both kinds of tissue and becomes a complete bundle. Connecting the cambium arcs of the adjacent bundles a cambium arc (z) forms in the intervening medullary ray, and this produces secondary medullary ray tissue on both its inner and its outer face. There is thus formed a continuous cylinder of cambium (x, y, z), though a somewhat irregular and wavy cylinder, standing between the zone formed within by the primary and secondary xylem-bundles and their intervening portions of the medullary rays, and the outer primary and secondary phloem-bundles with their intervening portions of the medullary rays. Although this cambium forms a cylinder, as stated, it is usually referred to as the "Cambium-ring," or "Cambium-circle," because it presents this appearance in transverse section. Provision is now made for the growth of all portions of the stele. Additional complete fibro-vascular bundles are now developed in the medullary ray spaces between the others, fed by a portion of the cambium in a similar manner. New medullary rays also develop in the substance of the bundles. We thus have developed upon the inside of the cambium-cylinder a cylinder of xylem, solid except for the blades of medullary ray tissue penetrating it nearly to the centre, and outside of the cambium-cylinder a hollow cylinder of phloem tissue or bast tissue, continuous except for similar but of course much shorter medullary rays. It has been said above that the portions of the cambium-circle opposite to the primary wood-bundles "may" produce secondary wood upon the inner face and secondary phloem upon their outer. While this does take place in some roots, it usually does not, only medullary ray tissue forming at those points on both the inner and outer

faces of the cambium. This constitutes the secondary structure of the root-stele, and any further growth which may occur is merely a continuation of the process described as secondary growth. When an annual resting period in growth occurs the ducts of the xylem produced toward the close of the year's growth will be conspicuously smaller than those produced at the beginning, so that conspicuous Annual Rings are produced in many woods.

After a tree has attained a certain age the wood at the centre dies, and becomes dryer and harder and of a different color from the living wood outside of it, and this dead portion becomes thicker year by year. It is called the Duramen, or "Heart-wood;" the outer is called the Alburnum, or "Sap-wood." It is the duramen only which yields the most of our colored cabinet lumber.

The effect of secondary growth upon the structures external to the bast cylinder is extremely variable, according to the extent of such growth and the relations of the phellogen and its structures and the individual habit of the plant. It has been stated that the phellogen may develop in any part of the cortex. It may now be stated that it may, and, in fact, usually does, in the root, develop in the bast cylinder itself, so that all the parts external to it, and even portions of itself, will belong to the periderm, or in the rare case of Bork-casting by the root will be cast off.

In all the classes which yield our medicinal roots the branches start from the pericycle outside of a xylem-bundle at the point h in Fig 353, as it is first developing and grows through the surrounding tissue to the surface. If a root-section has passed through branches these will appear upon the older part as mature secondary roots, which are successively less developed downward, appearing at length upon the younger portion as not having made their way through the overlying tissues to the surface. As the root first formed is called the Primary, so its branches are called Secondary. Their structural development is a repetition of that of the primary.

The continuity of growth in the root is uniform—that is, there is no division of it into joints or phytomers. There are hence no regular distances at which it branches, and when buds are produced upon it, as they are in rare cases, their points of origin are not so regulated.

STRUCTURE OF THE STEM.

(The following account of stem structure refers only to the ordinary plants of the flowering class. At its close a brief reference will be made to such others as require attention for the purposes of pharmacognosy.)

The history of stem development is best presented by contrasting it with that of the root, which has already been given. The three elementary tissues, dermatogen, periblem and plerom, are also found in the young stem structure. The epidermis and other tissues of the stem are more variable than the corresponding tissues of the root, and the details pertain for the most part to histology and to the special treatment of species or groups. The most important distinctions between the epidermis of root and stem may be mentioned as the presence in the latter of stomata. There is no extra development from the dermatogen at the tip corresponding to the root-cap, nor of hairs similarly aggregated to those of the root, although hairs of many forms abound upon the epidermis of the stem. Stem epidermis may consist of one or of several layers, and if of the latter they may be dissimilar in varying degrees. It may be persistent or it may suffer the same fate through the growth of the parts within it, which has already been considered in the case of the root.

The periblem of the stem develops structures in general similar to those from the root periblem, the most important distinction being the production of a chlorophyll-layer. A primary cortex, usually somewhat thinner than that of the root, is bounded externally by a hypoderm and internally by an endoderm, and may develop tubes similar to those mentioned as frequently pertaining to the root cortex, but, as in that case, no true vascular bundles. The effects of growth within the primary cortex of the root, leading to the formation and casting off of bork, we have seen to be of rare occurrence. In the case of the stem, however, it is of very common occurrence, so that the entire account which has been given of the development and disposition of periderm and phelloderm may be applied with especial force in the case of the stem.

The principal differences between root structure and stem structure are found in the developments from the plerom. Although, with the single exception among important medicinal stems of the male fern, there is but a single stele, in the form of a central cylinder, yet the development of its structure is markedly different from that of the root. Leaving out of consideration exceptions which are unimportant in pharmacognosy, we find that two distinct types of structure characterize respectively the monocotyledons and the dicotyledons and gymnosperms. The form characterizing the latter two will be first considered.

Vascular bundles originate in the plerom in the form of a circle, just as in the case of the root, the important difference being that each bundle consists, even in its primary state, of both phloem and xylem, with a cambium between. The typical form is that which in the root constitutes the secondary structure—namely, a bundle consisting of xylem within and phloem without the cambium arc, and this constitutes what is known as the Open Collateral Bundle. Secondary growth here consists in the addition by the cambium to each kind of tissue, and, in almost all cases, the development of new intermediate bundles and new medullary rays, as has been described in the case of the root. The result is that the general plan of structure attained is identical with that already recorded as ultimately attained by the most highly developed woody roots. There are, however, several differences which must be noticed. The most important is that the primary xylem bundles do not progress toward and meet one another at the centre, so that there is always left there a cylinder of the fundamental tissue, constituting the Medulla or Pith, which is connected through the primary medullary rays with the pericycle, or, after the disappearance of that and of the endodermis, with the cortex. The whole structure in transverse section may now be roughly compared with the wheel of a wagon. The pith corresponds to the hub, the primary medullary rays with the spokes, the spaces between the spokes to the primary wood wedges, the felloe to the bast product, except that the spokes should be seen extending through it, and the tire to the periderm in its various forms of development. Although the details of tissue arrangement pertain to histology, yet the deviations from the above relative positions of the phloem and xylem are of such very great importance in pharmacognosy that they are here referred to. We may have (1) the Bicollateral Bundle, in which a second fascicle of phloem is placed upon the inner face of the xylem, (2) the peculiarities characterizing the monocotyledons, which will be described later.

There are three ways in which the structure of the root or stem may be examined.

1. A Radial Section is a longitudinal section in a plane passing through the centre.

2. A Tangential Section is a longitudinal section in a plane which does not pass through the centre.

3. A Transverse Section is one passing exactly at right angles to the former two.

The appearance presented by a radial section through a perfectly developed woody stem possessing open collateral bundles may now be described as follows, enumerating the structures upon either side from the centre outward. (a) pith, (b) wood wedges, with medullary rays, the latter, if primary, communicating with the pith at the centre and outward with the cortex, and extending upward and downward from one node to the next; if secondary, extending outward and inward only through the growth of a single year—that is, through one annual ring—and upward and downward much less than the length of an internode; (c) the cambium, (d) the bast bundles, separated by their medullary rays; (e) the phelloderm, phellogen and periderm, the relations of which to one another and to the bast, and the structure of which, cannot be specified, owing to the extreme variation which they display in different stems. The composition of the bork, if

any, will also depend upon the point of development of the phellogen and its form upon the form of the latter.

The Bark is everything external to the cambium. It has been proposed to remove the word "bark" from common language, or to ignore its fixed common meaning, and to convert it into a technical name for the bork. Experience with English speaking people leaves no hope that they will consent to give up a word employed so widely and in such important ways, and its technical use can apparently result only in the introduction of a confusion, which is more wisely avoided by the coining of some new name, if that of bork is seriously objectionable, which does not appear to be the case.

Upon a transverse section the same structures as above recorded will appear, but instead of being in the form of thin strips upon either side of the centre they will be in the form of concentric rings around it. Thus the centre is seen occupied by a circle of pith, outside of which is a zone of xylem or wood tissue, separated by long or short medullary rays into its primary and younger wood bundles. Outside of the first annual ring is where the intermediate or secondary bundles make their first appearance. The secondary medullary rays (Fig. 355a) will be found not to extend outward or inward beyond the production of tissue of that year. Instead of appearing as blades, as they did in the radial section (b), the medullary rays now appear as narrow lines. That is, we now see the edges of the blades whose sides were before seen. Passing outward beyond the last of the annual rings, which successively exhibit a greater number of wood-bundles and medullary rays, we reach the cambium-ring. Outside of this we find the phloem or bast bundles separated by medullary rays continuous with those of the wood cylinder, and still outside of this the periderm.

The appearance of a tangential section will depend, of course, upon the tissues through which it passes. If it cuts the medullary rays these will appear neither as the broad sides, as at b, nor the edges of blades, as at a, but as transverse sections of them, as at c. If the ray consists of but one row of cells in width, then such a row will be exhibited upon the tangential section, its vertical height varying from a very few to quite a large number of cells. If, upon the other hand, it possess a lateral breadth of several thicknesses of cells, of 5 in our figure, this condition will exist only at its middle portions. At its upper and lower limits it will always be reduced to the thickness of a single cell, so that the tangential aspect of a medullary ray is almost always that of an ellipse, broad or narrow, according to the number of rows of cells of which it consists, in contrast with the extent of its upward and downward extension.

The pith or medulla in some stems after a time disappears more or less completely, leaving a cylindrical hollow cavity. This may be continuous through the nodes or separated at those points by transverse partitions.

In monocotyledons we have the Concentric Bundle, in which the one element surrounds and encloses the other. In all medicinal stems possessing concentric bundles, except the male fern, it is the xylem which encloses the phloem. If the two cylinders thus formed have not a common centre, the bundle is designated as Closed Collateral, and this is far more common than the typically concentric form. It is clear that in the last two forms a cambium cylinder, such as distinguishes the stele, possessing the form previously considered, cannot be formed. In such plants indefinite growth in thickness of the bundles obviously cannot occur, and the same is true of the entire stele, unless new bundles may develop in it. Usually this occurs not, but if the upper portion of the plant shall branch and continue to extend its leafy surface, meristem tissue will then form toward the outer portion of the stele, and from this new bundles will successively arise, so that the thickness of the trunk will keep pace with the extension of the crown, notwithstanding that the individual bundles do not increase in thickness after the completion of their primary structure. In stems possessing this form

of bundles the latter are found more or less scattered through the fundamental or medullary tissue, though there is commonly more or less of a concentration of them in some one region, usually toward the periphery of the stele. The endodermis of such plants is commonly known as a Nucleus Sheath.

Finally, we note that in many plants, represented among drugs by the ferns, the stele, as the stem, possesses a number, usually definite for the species, of vascular bundles, each invested by its own endodermis. In such plants no epidermis is developed, the hypoderm, developed from the periblem, being superficial.

VERTICAL AND LATERAL EXTENSION OF THE STEM, AND OUTGROWTHS FROM IT.

Examining a radial section of the tip of the stem we find, in addition to the structures already considered as belonging primarily to itself, protuberances, consisting of masses of meristem tissue belonging to the periblem and the dermatogen. Shortly each of these tissue masses assumes the condition of the primary growing point of the main stem. It may develop into a leaf, the structure of which will be considered further on, or a branch, which latter process is a mere repetition of that already considered in relation to the primary stem. In either case the vascular bundles exhibit a connection, variable in its details, with those of the stem from which it develops. The normal method is for a branch and leaf to develop together, the former in the axil of the latter, as already recorded. If two or more leaves, with their branches, develop at the same node, it results in the opposite or verticillate arrangement. If but one, then, of those developing at different levels, each is successively separated from the former by a uniform portion of the stem circumference, so that a spiral arrangement results. This spiral will be considered when we come to the study of the leaf.

The point at which one or more leaves develop has already been defined as the node, and the portion of stem intervening between two nodes as the internode. At first the internodes are so short as to be scarcely perceptible, but they continue to grow until a length more or less definite for the species is attained, so that the leaves and branches become separated by uniform vertical as well as circumferential spaces. This brings us to another great distinction between the stem and the root, in which latter we have found a continuous and uniform longitudinal development. The rule that a branch develops in each leaf-axil is habitually departed from in the leaf-representatives constituting the flower, and accidentally in many other cases. Its failure to develop may be temporary, although often very long continued, or it may be permanent. Upon the other hand the subtending leaf may fail, accidentally, or, in a few cases, habitually, to develop, so that the branch does not show its axillary nature. Finally, we note that a branch may accidentally, or, in some cases habitually, develop from some other point than the leaf axil, or two or more may develop, at least partially, from one axil, either side by side or in a vertical row.

Not only may a lateral branch thus fail to develop, but the apical extension of the growing point may fail, accidentally or habitually, the growth being continued by means of one or more branches only. A stem so formed is called a Sympodium or Sympodial Stem. One in which the apical growth is continuous is called a Monopodium or Monopodial Stem. Sometimes instead of the lateral or terminal growing point failing altogether to develop, it may develop in the form of some entirely different body from an ordinary stem, its growth in such cases usually ceasing permanently with the perfection of such an organ. When either of these results attends the terminal point, the one or more lateral branches may retain their lateral or oblique direction, or, if one, it may become erect and take the position naturally belonging to the main stem. If then the latter exists in a modified form it will be distinguished from a modified branch by being on the opposite side of the stem from the leaf, instead of being subtended by such leaf (Fig. 356, a the main stem, b the branch). Besides modified or unmodified leaves or branches, stems may develop various other appendages. When these are merely super-

ficial they are called Trichomes. The characters of trichomes upon stems or leaves, particularly the latter, are of the utmost importance in diagnosis. Their study, however, save as to the surface-characters which they collectively produce, pertains to histology. The gross surface character so produced will be taken up in connection with the leaf. When appendages are of deeper origin they are called Outgrowths or Emergences. These may contain vascular tissue, connected with that of the stem. Outgrowths are, for the most part, in the form of spines, hooks (Fig. 357a), warts, suckers (Fig. 358a*), or grasping organs. Both trichomes and outgrowths may be regularly or irregularly disposed.

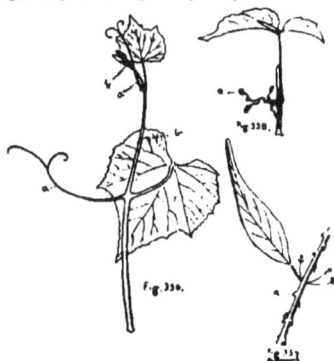

Roots may develop from branches which are subterranean, resting upon the surface of the ground or high above it. The latter may descend and enter the ground, fix themselves to a neighboring body for sustenance or support, or both, or extend into the atmosphere. They may even turn and enter a diseased or decaying portion of their own plant. They normally develop from the node only, but may develop from any other part or even from leaves.

An undeveloped stem or branch, or the partially developed summit of one, is a Bud or Gemma. The bud may be in a process of continuous development of its lower elements into mature phytomers, with the continuous production of a new growing point, or it may pass into a rest-

*In the species here illustrated the sucker is a stem, not an outgrowth.

ing state between successive seasons of growth. In the latter case it undergoes special modifications (Figs. 368, 371 and 372b). Its outer leaves become developed previous to the resting stage, but not as foliage leaves. They become modified instead in various directions as to form, proportions, relative position, appendages and exudations, to fulfil the office of protection as scales, and they subsequently fall away, never developing into foliage leaves. When no such provision is made the bud is commonly destroyed, with more or less of the young stem tip near it, during the resting period. Occasionally the bud is protected for a time by a special covering, formed by the petiole of the subtending leaf. It is then called a Subpetiolar Bud.

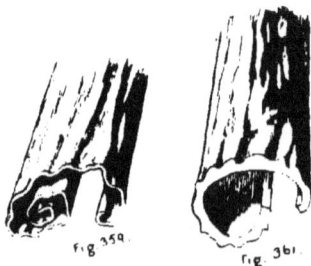

Fig. 359.   Fig. 361.

THE BARK.

Viewed from the standpoint of pharmacognosy, the bark, especially when detached from the remainder of the root or stem, is one of the most important portions of the plant. As has been seen, it is not a simple structure, but develops in part from the plerom, as well as from the periblem, and bears frequently, although this is not true of any detached medicinal bark, the epidermis as well.

In practice the bark is commonly differentiated into three layers; the Endophloeum, that portion resulting from the plerom; the Mesophloeum, the primary cortex, or the products of a phellogen developing external to the endophloeum, or both when they exist together; and the Exophloeum, consisting of a primary periderm. If, as is not the case in any medicinal bark, the epidermis persist, it will form the exophloeum. It has already been made sufficiently clear that a bark

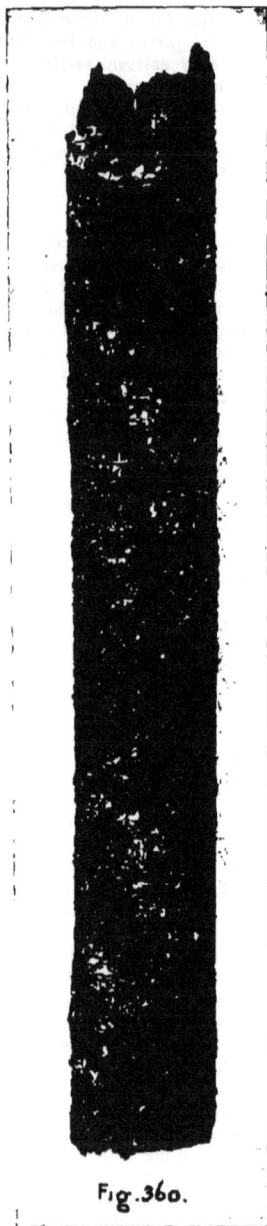

Fig. 360.

can come to want successively its exophloeum, mesophloeum, and even the outer part of its endophloeum. The study of barks includes a close examination of their cellular elements, as a preparation for which histological work is absolutely necessary. Examination of its gross characters involves, as the more important features, its extreme and average thickness, its manifest layers, as seen with a lens on transverse or radial section, their relative thickness, color, markings, consistency as shown by fracture, their separability from one another, that is, into laminae, together with the surface characters of the latter, the external color and level markings, the presence and nature of parasites, and the color and inequalities of the inner surface.

The laminae do not depend entirely upon different tissue composition. The same tissue, produced at different times, may present differences sufficient to result in different degrees of cohesion, as well as markedly different color, at different depths, so that separation may readily occur, or they may readily be distinguished in section. Groups or radial or tangential rows of tissue elements, differing from those adjoining, frequently produce gross markings on the section surface. The fracture of barks or of their individual layers is denominated in general as being brittle or tough. Various modifications are soft, earthy, granular, horny, waxy, fibrous, splintery or flexible. A bark may be flexible in one direction and not in another.

The outer surface is described in general as being harsh, rough, downy, smooth or shiny, and its lustre may be waxy, vitreous, and so on. Some of the elements causing roughness may require microscopical examination for their demonstration, while others are otherwise manifest. Care must be taken to distinguish between ridging and furrowing of other kinds. One kind is caused by a longitudinal wrinkling in drying, as in young Calisaya (Fig. 359). Another is owing to transverse (as in old Calisaya) or longitudinal (in the same) fissuring (Fig. 360). Another is caused by the elevation of corky ridges, or rows of corky warts, which may or may not become confluent in variable degree (as in Succirubra, Fig. 361). Fissures may characteristically open in the crest of a ridge or in the otherwise unchanged surface. Most color variegations are due to lichens or other parasites, and those due to lenticels are also very common.

A single color or shade of color of the inner surface is rarely characteristic, as it changes very greatly with age in keeping; but a carefully arranged series of them may be made diagnostic in many cases. The important characteristics of the inner surface depend upon the projecting bast bundles and contracting medullary rays. Very rarely indeed is the surface so free from these inequalities that it can be properly described as smooth. The slightest manifestation of the bundles gives the Striate condition. The striae must be examined as to length, straightness, direction as contrasted with the axis of the bark, apparent interconnection at the end, width, elevation and sharpness, with the complementary characters of the intervening furrows or pits. Some barks show a tendency to separate into laminae which run obliquely out upon the inner face, appearing there as partially separated tongue-shaped splinters.

CLASSIFICATION OF ROOTS.

Roots may be classified as to their duration, their order in time of development, place or nature of origin, function, form and consistency.

As to duration we have roots divided into two great classes, although the terms designating them are in general applied to the plant as a whole rather than to the root. Monocarpous plants are those which die after producing one crop of fruit, Polycarpous those which produce successive crops. The former are Annual when they live but a single season—as the rag weed and the sunflower; Biennial when they devote the first season to the storing up in some receptacle, such as fleshy root or bud, a supply of nutriment and fruit in the second season. The term winter-annuals has been applied to those which begin their life during the latter part of the first season, fruiting early the next season, so that their combined life during the two seasons is less than twelve months, as in the case of wheat and rye. Such may, by being planted early in the season, finish their existence during one season, as in the case of spring wheat. Those monocarpous roots which devote a number of years to the preparation for fruiting, as in the case of the century plant, belong to the Perennials. All Polycarpous roots belong, of course, to the perennials.

As to their order in time of development, the first root developing from the radical is the primary. All subsequently developed, whether from root or stem, are secondary, although those developing from secondary roots are sometimes designated Tertiary and so on. If the primary root continue its development so as to constitute a branch-bearing axis, it is called a Main-root or Tap-root (Fig. 362). If, instead of so doing, it divide at once into a number of approximately equal branches it constitutes the so-called Multiple Primary Root. This term has, however, been applied to a number of root clusters of similar appearance, but of very dissimilar origin. In some cases the primary root continues its vertical growth but does not increase in thickness to any appreciable extent. A number of similar roots then develop near its point of origin so that a fascicle of similar roots at length results, as in the onion. In other cases the stem of the plant takes root from one of its nodes, the portion below this point (Fig. 363a), with the original roots, perishing. To the cluster of roots thus resulting, although they are really secondary, the term "multiple primary" has also been applied. A true multiple primary root is of rare occurrence and does not exist among drugs. All roots which are not primary, or branches thereof, and all branches of roots which are not developed in regular order of succession, are called Adventitious.

As to their place or nature of origin, roots are Subterranean when they originate from points underground, whether from root or stem, and Aerial when they originate from points above the surface, whether from root or stem. A root may originate from an aerial point and afterward fix itself in the earth, as the Brace-roots of maize. A number of approximately equal and similar roots occurring in a cluster, especially if they be fleshy thickened, are denominated Fascicled. Roots existing in the form of a mass of thin, fibre-like, approximately equal and similar elements are called Fibrous.

As to their function roots are known as Absorbing, Fixing or Storage roots. A root of one kind may give origin to a branch of a different kind. Absorbing roots of parasitical plants are frequently greatly modified in structure to form Haustoria. Fixing roots are usually designated as Rhizoids. Storage roots are usually much enlarged and possess a fleshy consistency and characteristic forms. When only a limited portion of a root is fleshy-thickened, so as superficially to resemble a tuber, it is called a Tubercle. Care should be taken not to confuse this technical meaning of the term with its common use as designating a small tuber.

As to their form, roots are Simple when they do not branch, or Branched, Cylindrical, Terete (which includes the cylindrical and that form which differs only in that it tapers), Tapering, Napiform, when taking the form of a short, broad turnip (Fig. 364), Fusiform when spindle-shaped as some radishes (Fig. 365), Conical, or Cone-

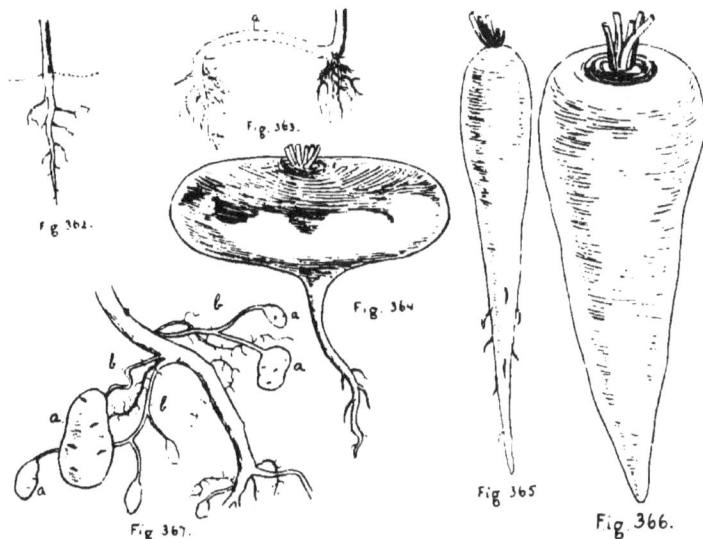

Fig. 363.

Fig. 364.

Fig. 361.

Fig. 365

Fig. 366.

Fig. 367.

shaped (Fig. 366), Capillary when very thin, long and hair-like.

As to consistency, they are denominated as Woody and Fleshy. By "fleshy" or "non-woody" we do not mean that wood tissue is entirely lacking, but rather that the proportion of the cellular, parenchym-atic or fleshy elements is so much greater than that of the woody that a woody character is not apparent. In practical pharmacognosy, where dried roots are mostly observed, a number of other terms, as in the case of the bark, come into use.

CLASSIFICATION OF STEMS, INCLUDING BUDS.

Stems may be classified as to duration, order of development in time, position and nature of origin, mode of extension, direction of growth and nature of support, modification of form or function and consistence.

As to duration they are, like roots, Annual, Biennial and Perennial. Annual stems are those which die at the close of the season. They may or may not pertain to annual roots. Plants possessing them are called Herbs. Herbs are therefore either Annual, Biennial or Perennial, in accordance with the character of the root. The definition of an herb is a plant, the aerial portion of the stem of which dies at the season's close. The stem of an herb is denominated Herbaceous. Biennial stems are those which are produced, usually underground like that of the po-tato (Fig. 367b), during one season, and perish after the production of their branches in the following season. Occasionally, however, like the cabbage, a biennial stem is aerial. Fleshy-thickened and biennial portions of underground stems, such as the potato, are denominated Tubers (Fig. 367a). Fleshy-thickened basal portions of stems which are invested by more or less fleshy-thickened storage leaves are called Bulbs. The varieties of the bulb will be considered under the subject of buds. Perennial stems are those which live and extend their growth from year to year. They are Determinate when their growth of the season is self-limited and closes with the production of a specially prepared Winter-bud, which protects the growing point for continued growth the next season; Indeterminate when no such bud is formed, growth continuing until the apical portion is destroyed by an inclement season. In such stems then we have the anomaly of a perennial stem with an annual tip.

As to their order of development in time stems are Primary, Secondary and so on, terms which are self-explanatory.

As to their position and nature of origin stems are Aerial or Subterranean, which terms are also self-explanatory. A secondary stem assuming an erect position from the base of the primary is a Sucker.

rooting at some of its joints, is called a Runner (Fig. 374). A horizontal underground stem, fleshy-thickened at least during the first year so as to serve as a storage receptacle, and giving origin to an aerial summit or branch, is a Rhizome (Figs. 368, 369, 371 and 372). The growth and duration of a rhizome may be indefinite, like those of stems, as in the case of the rhizome of Podophyllum (Fig. 369), or they may be restricted to one or to a definite number of years, after which the oldest existing phytomer perishes each year

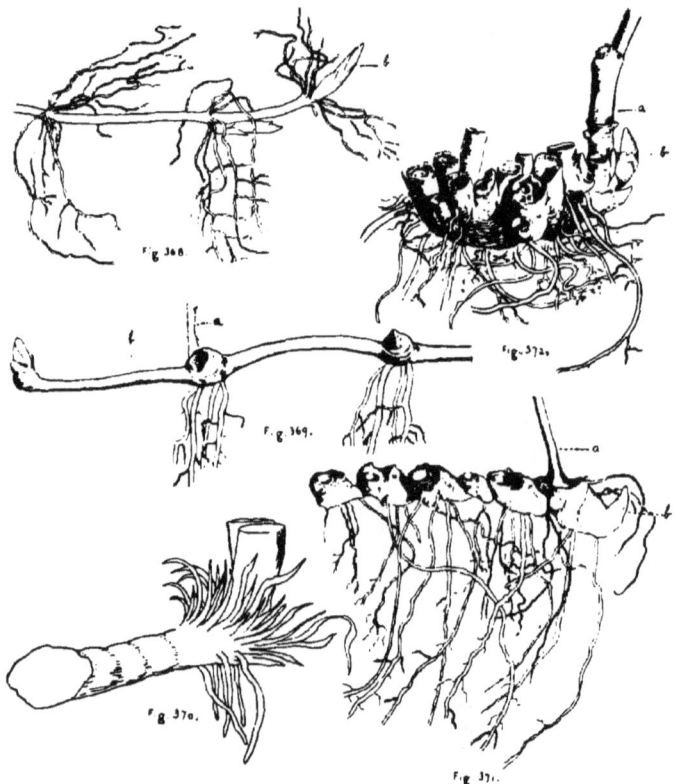

Fig. 368. Fig. 369. Fig. 370. Fig. 371. Fig. 372.

Such a one arising from a rhizome at a considerable distance from the original erect stem is called a Stolon. A short secondary stem developing from the base of the primary is called an Offset. An elongated, slender one, lying prostrate and

as a successive apical one is formed (Fig. 371). Rhizomes are so numerous and important in pharmacy that their characters call for special attention. They are classed as short or elongated, the former term referring to those the extreme length or shortness of which fall within certain fairly defined and restricted limits; the latter, those which either possess an indefinite extension, or the definite length of which is a great many times their thickness. Terms indicative of their form and consistency do not differ materially from those applied to other stems and roots. They are almost always sympodial. They are very subject to flattening, the flattened surfaces usually looking upward and downward. The presence or absence of branches is always characteristic. The manner in which the roots take their origin is equally so. These may form a circle (Fig. 368) or be restricted to the under surface (Fig. 369). The number of roots developing from a node is usually fairly characteristic. So is the persistency or brittleness of these, and the characters of the stumps or scars which they leave, as well as their form, which is very often triangular or quadrangular in section. Their structure, as observed either with the lens or with the microscope, is characteristic and of diagnostic value. Sometimes the roots are not only restricted to a certain portion of the node, but in the case of short rhizomes are restricted to a definite portion of the latter (Fig. 370). The relative length of the nodes of a rhizome calls for close attention, as compared with its diameter or thickness, and so does the absolute or measured length. The relations of the erect portions of the rhizome to the horizontal and the stumps or scars left by the former upon their death or separation constitute one of their most important diagnostic characteristics. Commonly disarticulation occurs with the production of a cup-shaped scar. This scar will be characteristic as to whether it form a depression in the general surface (Fig. 371) or be elevated upon a base (Fig. 372), as will the length of the latter, the form and depth of the scar and the characters of its edge. The size of the scar, that is its lateral breadth as compared with the thickness of the internode, is also noteworthy. Leaf scars, or leaf remains, upon rhizomes, call for the same examination as do the stem scars. They may surround the entire rhizome, in which case they are designated Annular, or they may be con-

fined to the upper surface. If the latter, the scar may be of characteristic form, as linear, elliptical, circular, cordate, crescent shaped or V-shaped (Fig. 373a). Finally we note that annular or longitudinal folds, thickenings, wrinkles or constrictions are characteristic of certain rhizomes as well as of roots, particularly in the dried state.

As to their mode of extension, stems may be Simple, or Branched. A stem denominated as simple is not necessarily entirely destitute of branches, as floral branches or small branches near the summit are permitted. It has already been shown that stems may develop monopodially or sympodially. The stem of a tree, which continues, except in case of accident, to develop monopodially, is called Excurrent. One which after a time loses its main stem in a number of branches, as for instance the elm, is Deliquescent.

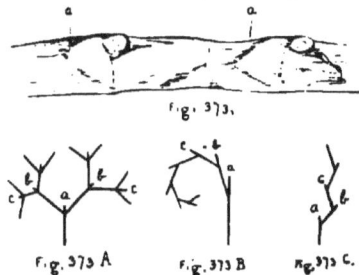

Fig. 373.

Fig. 373 A   Fig. 373 B   Fig. 373 C.

Sympodial stems possess several well-marked classes. If extension is by means of a pair of branches at the summit, these and their branches successively forking in a similar manner (Fig. 373A), the growth is Dichotomous or Bifurcating. (It is to be noted that these two terms are applied in a restricted sense to the form of dichotomy which is produced by the vertical division of an apical cell, so that the form figured in 373A is by some authors denominated "Falsely Dichotomous."). If but one branch develop from a node, the resulting sympodium may be of either of two classes, according to whether the successively developing branches are all developed upon one side of the axis (Fig. 373b), or produce the same effect by assuming such a uniform direction, in either of which cases we get the Secund form, or by developing alternately upon different sides (Fig. 373c) they maintain continuously the general direction of the axis, although it may be in an irregular or sinuous manner. In Figs. 369, 371 and 372, the apex of the stem, a, may be seen to each year become erect, producing the aerial stem of that year, the sympodial extension being by means of the branch, b, which will the next season in turn become erect and give origin to another horizontal branch.

The term Acaulescent, while meaning strictly stemless, can, of course, have no such application, as all flowering plants possess a stem even before germination occurs. The term is applied to those plants whose stems are so short as not to become conspicuous. The stem of such a plant is called a Crown. The term crown is also applied to the branching or leafy portion of any stem. A plant possessing a woody and erect stem rising singly to the height of fifteen feet or more is denominated a Tree, or Arborescent plant, although the precise application of such a term is impossible. A perennial woody stem which has not these characters is called a Shrub or a Fruticose stem. Very small shrubs appearing on casual inspection as herbs are called Undershrubs or Suffruticose plants.

As to the direction of their growth and the nature of their support, stems may be Erect, in which case they are erect through their entire length; Ascending, in which case the base for a greater or less distance rests upon the ground, the terminal portion becoming erect; Horizontal, in which case they are considered as having no other support than the parent stem, from which they extend at a right angle; Drooping, in which case they are first horizontal, the outer portion becoming pendant; Pendant, or "Weeping," when they are pendulous from their point of origin or almost therefrom; Decumbent, when at first erect or supported by the parent, the outer portion declined so far as to rest upon the ground; Reclining, when resting upon some means of support elevated above the earth, as over the tops or branches of other plants; Procumbent, when resting at full length upon the ground without

rooting at the joints; Repent, or "Creeping," when prostrate and rooting at the joints (Fig. 374); Twining, when supporting themselves by the twining of the stem itself around a support; Climbing, when elevating and supporting themselves by other methods than a twining habit, the principal forms being the Cirrhiferous, when climbing by tendrils (Fig. 356), and Aculeate, when climbing by hooks (Fig. 357).

As to modifications of form of function, stems are subject to a somewhat elaborate classification.

They may be modified for the purpose of defence, that is into thorns or spines (Fig. 375), although not all thorns or spines are transformed branches. Branches of this form sometimes remain so permanently, while at other times they afterward become foliaceous and develop into branches of the ordinary form (Fig. 376).

For the purpose of climbing they may become cirrose, that is, converted into Tendrils. The tendril may consist of the apex of the primary stem (Fig. 356), which then usually becomes forced to one side by the contiguous branch, which in turn fulfils the office of continuing the extension of the stem as a sympodium; or one of the branches may become the tendril. In the latter case the tendril will stand in the axil between leaf and stem; in the former it will stand upon the opposite side of the stem from the leaf (for tendrils connected with the leaf see following). A stem may instead become converted into a sucking disk, as in the case of Ampelopsis (Fig. 358). In this case the tip of the branch or stem becomes flattened and attaches itself very tightly to the supporting surface, so tightly, in fact, that a portion of stone or a splinter of wood may be torn from its support before the disk can be made to separate. Plants which grow in the water or in places subject to inundation may have portions of their stems inflated into bladdery forms to insure a floating condition. Such structures are, however, more commonly of a leafy nature.

Stems may become modified for the performance of the office of leaves. For this purpose the whole stem may become mod-

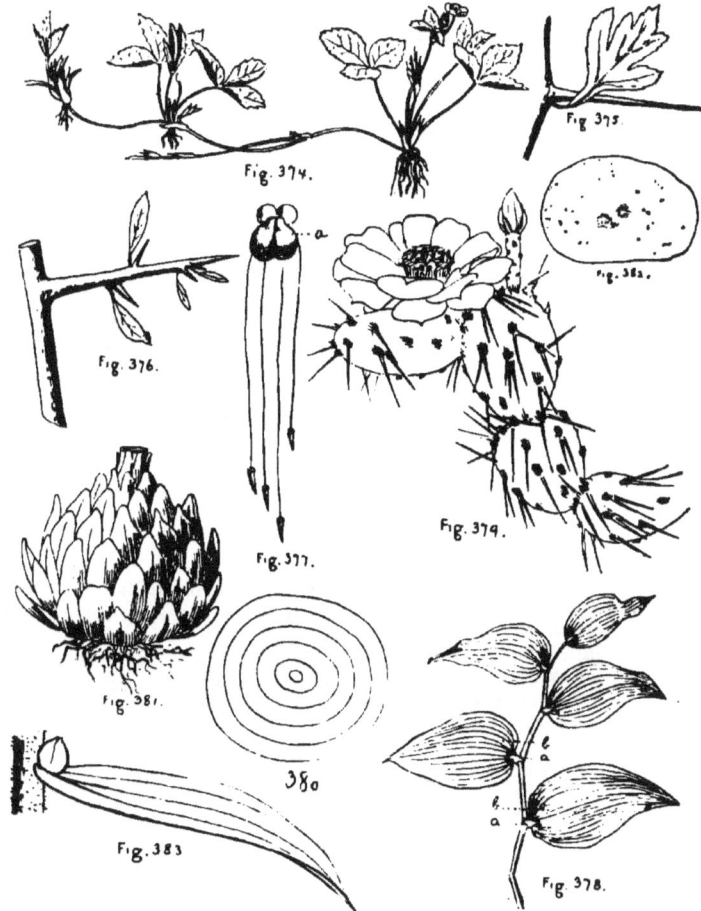

Fig. 374.

Fig. 375.

Fig. 376.

Fig. 377.

Fig. 382.

Fig. 379.

Fig. 381.

380

Fig. 383

Fig. 378.

ified into a single leaf-like organ, as in the case of certain aquatics, in which case it is known as a Frondose Stem (Fig. 377a). Upon the other hand, separate portions of the stem or separate branches thereof may become thus modified, as in the case of the so called "leaves" of the species of Asparagus cultivated as a decorative plant under the name of Smilax (Fig. 378; a, leaf; b, branch). Sometimes a stem or a joint of one, at the same time that it becomes modified to perform the office of a leaf performs the ordinary offices of a stem and important storage functions as well, as in the case of the Opuntia (Fig. 379). Such stems are called Consolidated. Besides such specially modified forms, a number of ordinary forms are characterized by the adjectives Terete, Cylindrical, Compressed, Triangular, Quadrangular, Alate or Winged, Costate or Ribbed, Chanelled, Striate and so on. In this connection the terms applicable to the superficial characters of barks already described, and those connected with leaf attachment, to be described further on, should be studied.

Besides the above stem-forms, which admit of ready classification, we have a large number of modifications to effect special purposes which must be considered individually. As these possess but a slight interest in relation to pharmacognosy we refer the interested student to general works on botany.

An important office of the stem is the storage of nutriment. All stems perform this office to a greater or less extent, but some are especially modified in form for the purpose. Of these we have already referred to special rhizomes and tubers. It remains, then, only to consider the various forms of the bud, including in this term all forms of the bulb. A bulb which, like the onion (Fig. 380) has its fleshy thickened leaves in the form of broad sheathing organs, seen upon transverse section in the form of concentric rings, is called Tunicated or Coated. Those like the lily (Fig. 381), in which these leaves appear in the form of narrower projecting scales, are called Scaly. When in the axils of the scales we find smaller or secondary bulbs or buds it is a Compound bulb. When the texture of a bulb is so dense that its leaf-elements are not conspicuous it is designated a Solid bulb. When it is still more dense, as in the case of the gladiolus (Fig. 382), so that the leaves are not to be distinguished by ordinary methods, it is a Corm. In some plants the axillary bulbs, instead of occurring in the axils of the bulb-leaves, occur higher up in the axils of the ordinary foliage leaves, as in the case of the tiger lily (Fig. 383). Their true nature as buds is in this case conspicuously shown, and they are sometimes spoken of as Bud-bulbs. In other related plants similar bulbs are densely aggregated in a terminal umbel looking like an inflorescence, as in some species of onion (Fig. 384).

Buds proper admit of an elaborate classification, which, although not of such interest in pharmacognosy as to warrant its study in this place, is of fundamental importance in systematic botany, and furnishes a key to many problems which are

otherwise very abstruse. The study of buds is called Vernation, and that of the arrangement of the leaves composing them Praefoliation. In general the ar-

Fig. 384.

rangement of leaves in the bud admits of the use of terms similar to those applied to the parts of the perigone in a similar state.

Buds may be classified as to their structural form, their position, and parts. A winter bud which protects itself by specially developed scales is known as a Scaly bud; one which does not, a Naked bud. A bud consisting only of leaves is a Leaf bud; one only of a flower, a Flower bud; one consisting of both, a Mixed bud. Solitary buds occurring in the axil of the leaf and developing at the regular time are called the Normal buds. Any buds in addition to the normal bud, occurring in the leaf axil, are called Supernumerary. They may be situated above or at the side of the normal bud. The normal bud is sometimes situated a little above the actual axil, in which case it is called Supra-axillary. All the buds so noted are denominated Lateral, in contradistinction to the single terminal bud, but it is to be noted particularly that buds lateral as to their origin may become terminal through the effects of sympodial growth. Buds which develop at other points than the apex or axil—as, for instance, from an internode, a leaf, or even from a root, as well as those of axillary origin, but developing out of their regular order—are called Adventitious. The latter form of adventitious buds, resulting from retarded development, are known as Latent buds.

THE LEAF.

It is possible to determine but very little of the leaf structure without the aid of a microscope. The gross parts of the leaf have already been defined (see Fig. 3) as the base, consisting chiefly of a pulvinus, the stipules, petiole and lamina. In addition to these we have to note that many leaves possess a basal portion which sheaths the stem, but which is not exactly in the nature of a petiole or of stipules, and is denominated the Sheath (Figs. 385 and 386a). Leaves

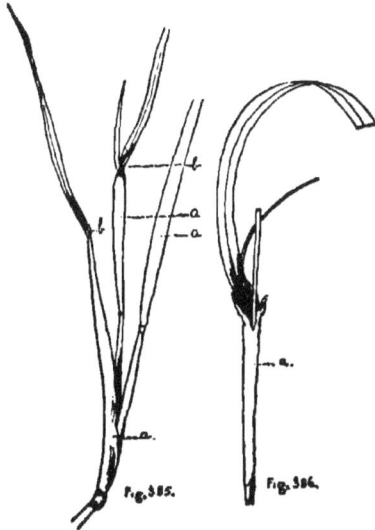

Fig. 385. Fig. 386.

possessing a sheath frequently bear at the point of its junction with the lamina a blade-like appendage, usually very small, and analogous, if not homologous, with the crown of a petal, denominated the Ligule (Fig. 385b). It has also been shown that the leaf originates and develops as an extension of the periblem, covered by that of the dermatogen, and that it develops a stele (or more than one) which becomes continuous with that of the stem. In other words, its mode of development is precisely like that of a stem-branch. We have in it, therefore, all the elements which characterize primary stem structure. Just as branches of the cauline stele pass into leaf and branch, so do those from the foliar stele

pass laterally into its expansions, and secondary and tertiary ones successively pass from the latter. These branches are very frequently joined at their distal ends to others (Fig. 388a), as well as at their proximal ends to the parent system. Whether such is or is not the case, the result of the branching is the production of a framework or skeleton which forms a support to the parenchymatous tissue which fills its meshes and covers its surface, being in turn covered by the epidermis. Bundles which obviously separate from one another at or near or below the base of the blade, and maintain their course well toward the apex or margin, are called Costae or Ribs if equally prominent (Fig. 474). Nerves if lateral and markedly less prominent than one or more of the central ones (Fig. 412).

The central one, whether there be others or not, is the Midrib (Fig. 387a). Branches of ribs or nerves are called Veins, and they are distinguished as Primaries (b) when departing from the midrib, Secondaries (c) when departing from primaries, and so on. Primaries of lateral ribs or nerves must be specially so designated in description. When there is connection at the distal ends (Anastomosis or Reticulation Proper, Fig. 388a) it is essential that the order (Primary, Secondary, and so on) of the anastomosing veins be stated, also the region of the leaf where the anastomosis takes place. Very small veins are called veinlets.

Except as to the general terms which follow under leaf classification, it is impossible to ascertain the structure of the cortex and epidermis of the leaf by ordinary methods, so that this subject is relegated entirely to the department of histology.

Very rarely has the leaf a terete form and a radial structure as seen in transverse section. Typically it is a flattened body. One flattened surface, the Ventral, faces upward or toward the stem which bears it and is ordinarily spoken of as the upper leaf surface. The under or outer surface is technically known as the Dorsal. By a partial twist in the petiole the surfaces may become laterally placed, the edges vertical. In a few leaves the surfaces are normally in the latter position. Between the dorsal and ventral surfaces there are usually differences sufficient to necessitate their description separately. In such descriptions it is better to speak of the dorsal surface as being underneath rather than "below," as the latter term may confuse it with the basal region. The connection of the leaf with the stem is usually by its base, consisting of a pulvinus and forming an articulation, the structure of which is designed to afford a prompt and ready separation of the leaf at the conclusion of the performance of its function, as well as to provide for certain movements and changes of position during life. In this case the stipule is often present as a prominent organ and of exceptional value and importance in classification and diagnosis, as in such families as Rubiaceae, Salicaceae and Violaceae. It is to be studied as to duration, falling sometimes before the expansion of the leaf, or persisting to various degrees after this period, and as to its freedom from or degree of adnation to the leaf, the stem or to the adjacent stipule in case of opposite leaves. In the latter case the organ formed by the union of two stipules belonging to opposite leaves is designated an Inᵗᵉⱼapetiolar stipule (Fig. 389a). Especial importance attaches to the characters of these, as they frequently develop into remarkably formed and appendaged bodies, as especially in the Rubiaceae, a most important medicinal family. Cases even occur in which the two stipules of a single leaf become contiguous and connate. The stipule is, moreover, to be studied in every detail in which the leaf is to be studied in connection with the classifications which follow. Leaves which do not possess stipules are called Exstipulate; those which do, Stipulate. Stipulate leaves are very frequently mistaken for exstipulate when the stipules are caducous or fugacious. It is frequently the stipules which are modified to form the scales of scaly buds.

At other times the connection of the leaf to the stem is by means of the adnation of more or less of its petiole, or even of its blade thereto, producing in the former case a ridge, in the latter a pair of herbaceous wings upon the stem (Fig. 395). In this case perfect articulation does not occur, and stipule characters are either wanting or greatly modified.

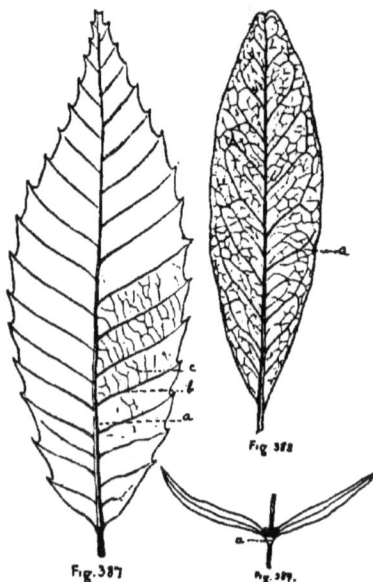

Fig. 388

Fig. 387.   Fig. 389.

The petiole instead of being adnate may be wanting, or it may be very short. In this case the relative position of the base of the blade to the stem will depend upon its form.

The attachment of the petiole to the blade is always really marginal, though by the cohesion of basal lobes (Fig. 390) it is often apparently intra-marginal or even central. Basal lobes may, upon the other hand, be adnate along the petiole, or the same appearance may be produced by the gradual differentiation of petiole into blade. When the margins of the petiole throughout are herbaceous and in continuation with the blade. the petiole is said to be Margined or Winged. In the blade of the leaf the cortex and epidermis bordering each branch system of a rib and its veins may be entirely continuous with that of the adjacent one upon either side, so that the blade will be entirely unsegmented, and its margin entire (Fig. 388), or, upon the other hand, cortex and epidermis may, at the end and for a variable distance toward or even to the base of its branch system, be separate from its neighbors, thus making a mar-

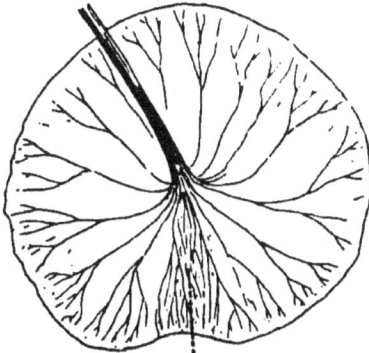

Fig. 390.

gin more or less toothed, lobed or divided. When the division is thus carried entirely to the point of origin of such branch system, the lamina of the latter may still be connected at its base with the midrib, or it may be entirely separate from that also, so that the connection of that branch system with the rest of the leaf will be by its principal vein only (Fig. 391, &c.). An articulation will then form there, and the lamina will consist of a number of distinct blades of secondary rank, thus giving us the Compound Leaf. These secondary blades may be similarly divided, giving us the Decompound Leaf. The divisions of a compound and decompound leaf are called Leaflets, and are subject to the same conditions, description, and classification as primary leaves. Decompound leaves are spoken of as once compound, twice compound, and so on. Stipules, called Stipellae (Fig. 391a) may develop at the base of leaflets, but buds do not form there, a distinction which is sometimes useful in determining whether a body is a branch, bearing simple leaves, or a compound leaf. The continuation of the petiole throughout the blade of a compound leaf, is called its Rhachis (Fig. 391b).

Leaves, considered in the gross, are to be classified as to duration, their retention upon the plant, texture, surface, attachment to the stem, attachment of the blade to the petiole, form, including general outline, special form of base and of apex, venation, margin, division, modification of form and function.

As to duration leaves are Annual and the trees producing them Deciduous, when their duration is through a single season only, and Evergreen, when they remain in their normal and active condition into the succeeding season. Evergreen leaves may be either biennial, the ordinary form, or perennial. Persistent leaves are those which remain upon the tree, but in a dead condition, until forced off by the growth of the following season.

As to their texture and consistency, the ordinary form of leaf, in which it possesses active chlorophyll tissue, is denominated Herbaceous, in contradistinction to the Scarious or Scariose form, in which

Fig. 391.

it has a dry and papery texture. Herbaceous leaves are Membranaceous in their ordinary form, that is, not excessively thickened, Coriaceous when tough and leathery, Fleshy or Succulent when largely parenchymatous, thickened and juicy. A leaf which exhibits translucent dots when held against a strong light is called Punctate. The surfaces of leaves may be classified in two ways, first, as to the characteristics of the individual trichomes (Indumentum) which they bear; second, as to the general surface effects which result from the latter. The former method, while not admissible in this article, is of very great importance in the characterization of medicinal herbs and leaves, especially as it constitutes one of the greatest aids to the identification of powders. The latter method can only be studied with advantage by the actual examination of type specimens, it being almost impossible to characterize the different forms by definition. A surface is Opaque when it is not shining or lustrous. It is Glabrous when it does not possess any trichomes in such forms as to detract from a smooth surface. It is Glaucous when covered with a waxy exudation, imparting to it a peculiar whitish appearance ("bloom"), such as characterizes the surface of an ordinary black grape. It is Scurfy when covered with more or less of an appendage in the form of granular or detached scaly masses. When the matter of such masses is more thinly distributed, appearing in the form of a powder rather than a scurf, the surface is called Pulverulent. A Pubescent surface is a hairy surface, which is not readily distinguished as pertaining to any one of the other specific classes. If the hairs of a pubescent surface are very short and fine, so that the consequent roughness is reduced to a minimum, the surface is called Puberulent. If a hairy indumentum is fine and of an ashy gray color, the hairs not arranged in any regular direction, the surface is Cinereous. If the hairs all lie in one direction, are closely appressed and have a shining or silky lustre, the surface is called Sericeous. If this lustre is intensified and of a strongly whitish color, whether the trichomes be hairs or scales, the surface is denominated Argenteous. Such hairs as are capable of producing a sericeous surface are themselves denominated sericeous or silky, even though they be in insufficient numbers to impart this character to the general surface. A surface tending toward the sericeous, but not sufficiently pronounced, is called Canescent. When there is a dense covering of more or less elongated and matted hairs, the surface is called Tomentose. When such a covering is thin, its hairs

less elongated, it is called Tomentellate. When there is a covering of thinly distributed, elongated, moderately soft hairs, which are not closely appressed, the surface is Pilose. When hairs are similarly distributed, but are elongated and coarse, the surface is Hirsute. When similar coarse hairs are rather stiff, lie

Fig. 392.

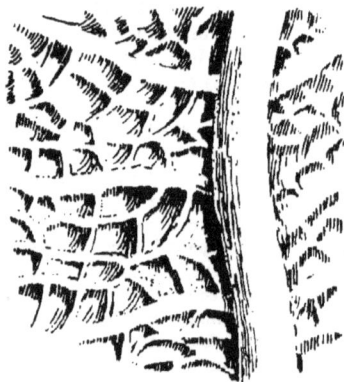

Fig. 393.

in one direction, somewhat appressed, and particularly when each develops from an elevated base, the surface is Strigose. A surface which possesses an indumentum of scales is called Lepidote. When the indumentum consists of hard, elevated points, giving a roughness to the surface, the latter is Scabrous. When such elevations are more pronounced, unyielding and sharp pointed, the surface is

Hispid. A surface which is roughened by the presence of numerous, closely set wrinkles is Rugose. When a surface is made up of small, blister-like elevations consisting of the arching interspaces between the veins it is Bullate (Fig. 392). The opposite surface, containing the cavities of the bullae, is called Cancellate (Fig. 393). When the hairy covering is chiefly confined to the margin, presenting itself in the form of a fringe of hairs, the term Ciliate is applied (Fig. 391). Finally it is to be noted whether the veins or ribs, and which of them, are prominent upon both sides or either side, or whether, upon the contrary, they are depressed (called Impressed) below the general surface. At times a rib or vein will not be impressed, but will yet be channelled, so as to appear impressed upon casual observation. A surface which is marked by spots, differing in color from the remainder of the surface, is called Maculate.

As to their attachment to the plant stem, leaves are Sessile (Fig. 394) when they possess no petiole. When a petiole or a lamina has grown fast for a portion of its length, to the plant stem, it is Adnate (Fig. 395). One whose base is heart-shaped and surrounds the plant stem, whether growing fast to it or not, is called Amplexicaul or Clasping (Fig. 396). When the basal lobes of a clasping leaf entirely surround the stem and become connate upon the other side, so that the stem appears to be growing up through a perforation in the leaf, the leaf is called Perfoliate (Fig. 397). When opposite leaves are connate by their bases they are called Connate or Connate-perfoliate (Fig. 398). When the base of the leaf has its ventral surface brought into contact with and surrounding the stem it is called Sheathing (Figs. 385 and 386). When the bases of sheathing leaves clasp the stem in such a manner as to present a V-shape in transverse section. and one is superposed upon another in the same manner, they are called Equitant. Leaf sheaths are divided into two classes, according to whether the edges are connate upon the opposite sides of the stem (Fig. 386), giving us the Closed Sheath, or

merely meet or overlap there without becoming connate (Fig. 385), giving us the Open Sheath.

As to the attachment of the blade to the petiole, the leaf is Peltate when this insertion is intra-marginal through the connation of the edges of basal lobes. A peltate leaf may be Centrally (Fig. 390) or Eccentrically (Fig. 399) peltate. When the petiole changes so gradually into the lamina that it is impossible to say where one begins and the other ends, we say they are Continuous (Fig. 400). When the margin of a blade is continued downward along the sides of a petiole, the latter is called Margined (Fig. 401). When these margins are manifestly herbaceous it is sometimes distinguished as being Winged, although this application of the term is not to be commended. Some of the terms here given connect this aspect of leaf classification with that which follows in reference to the distinctive form of the base.

By the Outline of the leaf, we refer to the general form of its margin, whether that be entire, or not. If not, then the general form of an outline is formed by connecting the extreme points of its margin with one another (Fig. 402, an obovate outline). It matters not, therefore, whether a leaf be entire, toothed, lobed or parted, or even if it be entirely compound or decompound, its outline will be the same provided a line connecting its extreme marginal points with one another possess a given form. The forms of leaves on this basis may be divided into three general classes—(a) those broadest at or about the middle, (b) those broadest at some point above the middle, (c) those broadest at some point below the middle. Of the first class, beginning with the narrowest, we have the Capillary or Hair-like forms, the Filiform or Thread-shaped (Fig. 403), the Acerose or Needle-shaped (Fig. 404), as those of the pine, and the Linear or Ribbon-shaped (Fig. 405), all of which are so elongated that they present the appearance of being about of uniform width throughout. A leaf similar to but shorter than the linear in proportion to its breadth, without regard to the character of its apex or

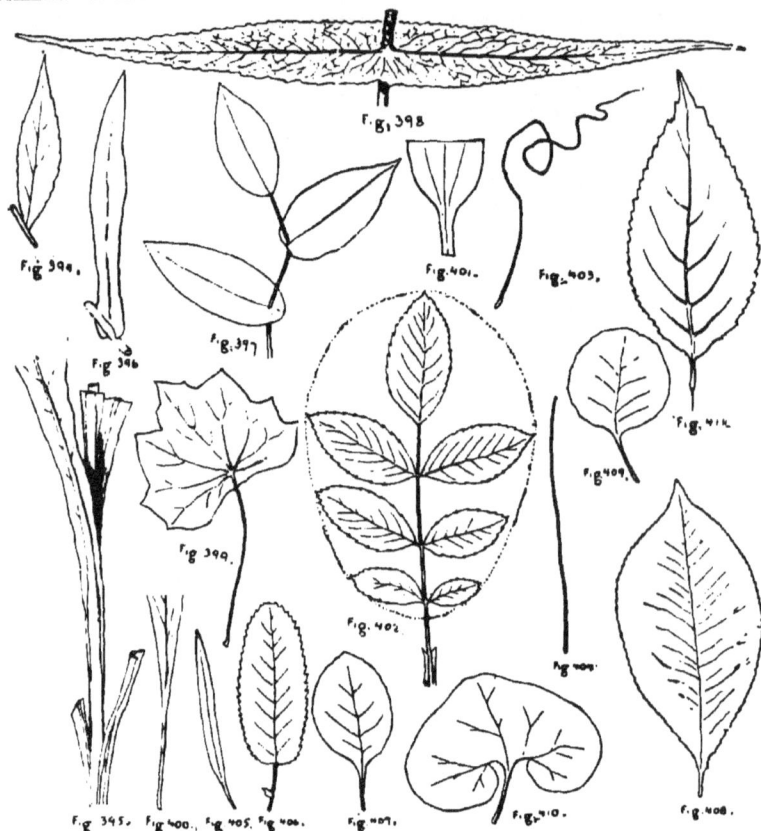

Fig. 398

Fig. 394. Fig. 401. Fig. 403.

Fig. 396 Fig. 397

Fig. 411.

Fig. 409.

Fig. 399.

Fig. 402

Fig. 345. Fig. 400. Fig. 405. Fig. 406. Fig. 407. Fig. 410. Fig. 408.

base, is Oblong (Fig. 406). One of similar form, but having a length not more
than twice or thrice its breadth, and narrower than a circle, is Oval (Figs. 407
and 408), a term which must not be con-
founded with Ovate. If an oblong or an
oval leaf possess a regularly rounded
outline into and through the apical and
basal portions it is called Elliptical. We
have, therefore, two forms of the elliptical leaf, denominated respectively Ob-
long-Elliptical (Fig. 406) and Oval-Ellip-
tical (Fig. 407). A circular leaf (Fig.
409) is called Rotund or Obicular. Finally we have the leaf which is broader
than circular—that is, its lateral diameter
is greater than its vertical, and this is
called Transversely Elliptical.

Those which are broadest at some point
below the middle or above the middle
should, in description, besides being designated by the class name of their form,
have it specified in some way as to about
the portion at which the greatest breadth
occurs. Beginning with the broadest
ones we have that which is broader than
long and with a heart-shaped base, called
Reniform (Fig. 410). One which possesses a length greater, but not more
than two or three times its breadth, is
called Ovate (Fig. 411). One of similar
form, but its comparative length greater,
is called Lanceolate (Fig. 412). One
which is ovate, but with the greatest
breadth at the very base, the margins not
or but little curved, so that it is approxi-

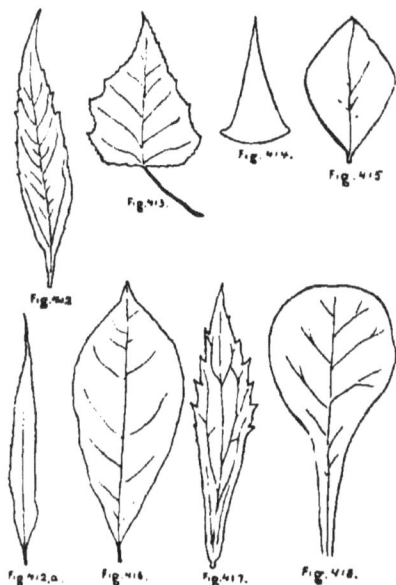

Fig. 414.

Fig. 415.

Fig. 413.

Fig. 412.

Fig. 412,a.

Fig. 416.

Fig. 417.

Fig. 418.

as regards the comparative length and breadth of the leaf are Sickle-shaped, Scimeter-shaped, and so on.

A large number of terms are employed to indicate especially the form of the apex of the leaf. Beginning with one which is inversely cordate—that is, with the sinus at the apex—we have the Obcordate form (Fig. 421). When the sinus is smaller it is called Emarginate (Fig. 422), and when very slight, Retuse (Fig. 423). If the sinus be an angular one with straight sides, it is called Notched (Fig. 424). If the apex be abruptly terminated, as though cut across in a straight line, it is called Truncate. If any portion of the apex of the leaf be narrowed into a point, the leaf is called Pointed (Fig. 425). If such narrowing be gradual, so that the point is considerably longer than broad, even though it be preceded by an abrupt narrowing, it is called Acuminate. If the acumination is preceded by an abrupt narrowing, then it

mately triangular, is called Deltoid (Fig. 413). One still narrower, but of similar form, bearing the same relation to the lanceolate which the deltoid does to the ovate, is called Subulate, or Awl-shaped (Fig. 414). An ovate or oval leaf whose outline instead of being regularly curved is made up of four comparatively straight lines, is called Trapezoidal or Angularly-ovate. Another term which is applied to it is Rhomboidal (Fig. 415).

Most of the forms just referred to are paralleled by exactly similar forms in which the widest portion is above the middle. The names for these are formed by prefixing the syllable *ob* to the corresponding name of the other form; as, Ob-ovate (Fig. 416), Oblanceolate (Fig. 417), and so on. When an Obovate or Ob-lanceolate leaf possesses a broad, rounded apex, and a somewhat elongated lower portion, it is called Spatulate (Fig. 418).

The outline of a leaf is greatly modified when the portion upon one side of the midrib is longer or broader than that upon the other, giving us Inequilateral, Unequal or Oblique forms (Fig. 419). When such a leaf has its midrib laterally curved it is styled Falcate or Sword-shaped (Fig. 420). Modifications of this

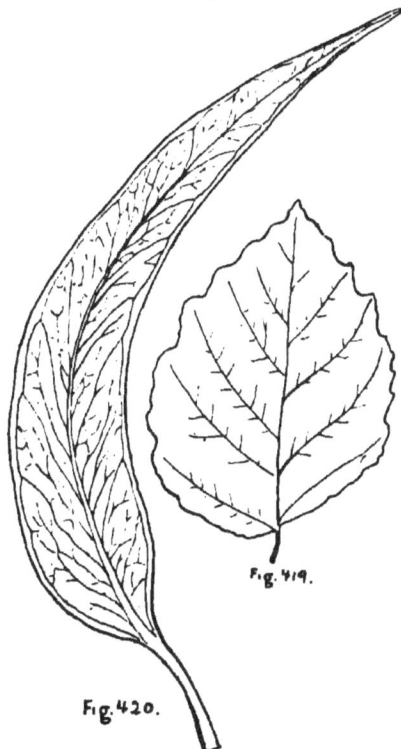

Fig. 419.

Fig. 420.

is distinguished as being Abruptly Acuminate (Figs. 426 and 427). If the narrowing be very gradual and not preceded by an abrupt narrowing, the apex is said to be Tapering (Fig. 428). If the point of the leaf be extremely abrupt and very small, it is Mucronate when soft and herbaceous, Cuspidate when hard and stiff, like a tooth. Any of the above mentioned forms may be either Acute, when the ultimate apex is sharp (Figs. 427, 428 and 429), Obtuse when not so (Figs. 425 and 426), Blunt when very obtuse (Fig. 430), or Rounded (Fig. 431). A leaf which has the midrib only extended into a bristle-shaped point is called Apiculate (Fig. 432), and this condition can apply to a cordate as well as to other forms of the apex.

The special forms of the base of the leaf-blade yield a correspondingly large number of terms. The terms cordate, truncate, rounded, blunt, obtuse, acute, acuminate, abruptly acuminate, require no additional definition to those which have been applied to similar forms of the

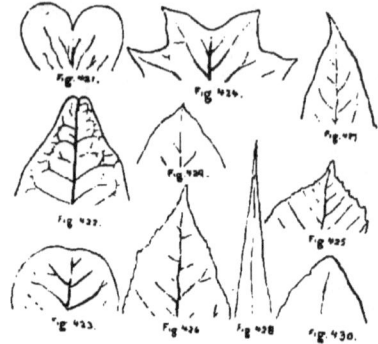

apex. When the two sides of the base are straight and come to an acute point the base is called Cuneate or Wedge-shaped (Fig. 433). A base which at first assumes a form which later yields to a sudden downward prolongation, or acumination, is called Produced. In all forms of the cordate base the greatest care must be taken to specify the precise character both of the sinus and the lobes. The for-

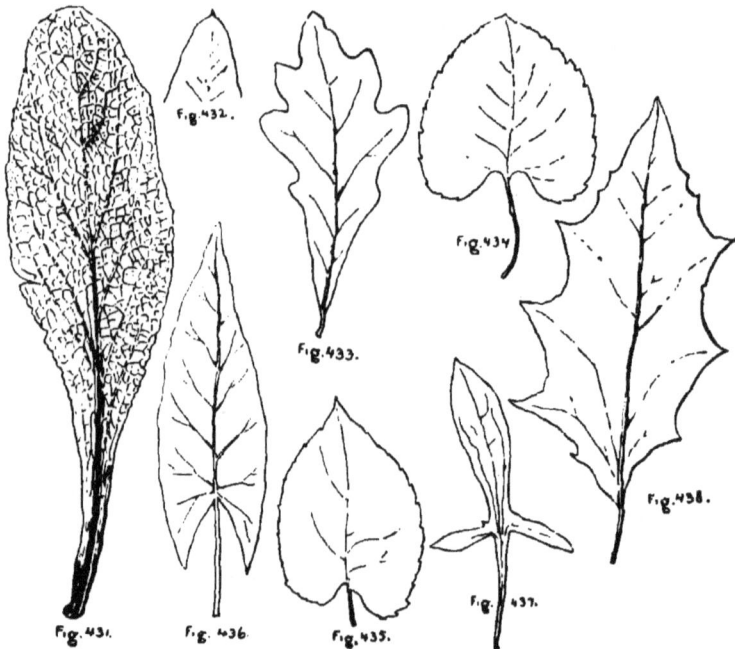

mer must have its form or outline specified as well as the angle which it makes. It should, moreover, be carefully noted whether the leaf base at the summit of the petiole be produced into the sinus, in which case it is called Intruded (Fig. 434). Sometimes the lobes of a cordate base will meet one another, or even overlap. The forms of the lobes are also capable of taking descriptive titles similar to those characterizing the lamina in general. The principal of such terms are Auriculate when the lobes are rounded similarly to the lobe of the human ear (Fig. 435), Sagitate when pointing downward and acute, like the lobes of an arrow head (Fig. 436): Hastate or Halberd-shaped when turned outward (Fig. 437). A base is Oblique or Inequilateral when descending lower upon one side than upon the other (Figs. 435 and 438).

The greatest importance in descriptive terminology pertains to the classification of leaf venation, owing to the frequency with which leaves must be identified in such a fragmentary state that there is little beyond the surface and venation, with possibly a portion of the margin, to assist us. The forms all fall within two principal classes, which in general characterize respectively the monocotyledons and the dicotyledons. The former bears its principal veins more or less parallel with one another, and these are numerous. Such leaves are called Parallel Veined (Fig. 439). In the second form there is but one, or a comparatively few, original veins, and these give rise to successively developed branch systems, the whole forming a network or Reticulum. Such leaves are called Reticulated or Netted veined (Fig. 387, &c.). These may or may not anastomose or intercommunicate at their distal ends. When they do, the term Reticulate is applied to them in a special or restricted sense (Figs. 388 and 421). In leaves of the last-named class the details of the method of intercommunicating are very important. Thus, in some cases, the end of each primary is arched upward into the next primary above (Fig. 388). In such case it is important to note the comparative distance from the margin at which the communication takes place and

the angle at which the two meet, as these characters are always constant in the same species. In other cases the primaries (or the ribs, as in Fig. 474) are directly connected by straight and parallel secondaries, or in still others (Fig. 421) by an irregular intervening network of small veins. Primaries connected by the first method are usually also connected near the base with the midrib by a number of curved secondaries. When the principal veins or nerves of a leaf are straight they are called Rectinerved, when curved, Curvinerved. The latter term refers to a regular and characteristic curve, not to a crooked course. Some leaves are characterized by possessing waving or crooked nerves or veins. Two great classes of netted veined leaves are recognized, the one in which there is a main Rhachis or midrib, from which primaries extend regularly toward the margin. This form is known as the Penni-nerved or Pinnately veined leaf (Fig. 387). The number of pairs of primaries, whether they originate exactly opposite to each other or somewhat irregularly, is within fair limits characteristic of the species and should be stated. The same is true of the angle at which they radiate from the midrib. In the case of additional ribs or nerves of such a leaf, the number and stoutness as compared with the midrib, their comparative length and the position which they take in the leaf are all important. The second great class of netted-veined leaves is that in which a number of approximately equal ribs radiate from the basal region. Such leaves are known as Palmately or Digitately Veined (Figs. 440 and 441). There are, of course, many forms of intergrading (Figs. 442 and 474) between such leaves and pinnately veined leaves with secondary ribs or nerves. Sometimes the nerves start from the very base of the leaf, in which it is called Basi-nerved (Fig. 440); at others from the lower portion of the midrib, when it is called Costi-nerved (Fig. 442). When the ribs or nerves are manifestly continued downward into the petiole, the leaf is called Flabellately nerved (Fig. 441).

The manner in which the leaf margin comes to deviate from an entire condition

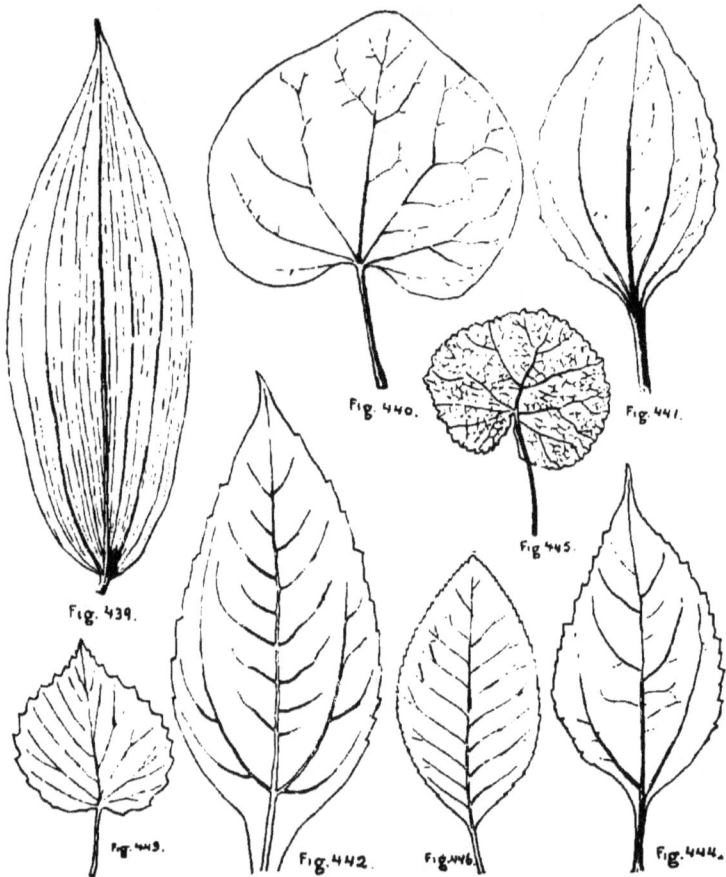

Fig. 440.

Fig. 441.

Fig. 445.

Fig. 439.

Fig. 443.

Fig. 442.

Fig. 446.

Fig. 444.

has already been indicated. Three special forms of toothing are recognized, in accordance with the form and direction of the teeth. When the latter point in an outward direction the margin is called Dentate (Fig. 443); when directed toward the apex of the leaf, Serrate (Fig. 444); when, instead of being pointed, the teeth are rounded, the margin is Crenate (Fig. 445). Diminutives of these terms, indicating that the teeth are very small, are Denticulate, Serrulate (Fig. 446) and Crenulate. To any of these terms the word "Minute" may be prefixed as indicating that the teeth are still smaller. Of each of these three principal forms there are a number of sub-forms. When the teeth bear smaller or secondary teeth, the word

"Doubly" is prefixed (Fig. 447, doubly serrate). When serrate teeth have their points very strongly directed toward the apex or appear as though pressed inward against the margin, they are called Appressed (Fig. 447, partly). They may even be Incurved. When, upon the other hand, the ends of the teeth are turned outward, they are called Salient. When the points of the teeth are very fine and produced in the form of bristles they are called Spinulose (Fig. 387). When a margin shows indications of being dentate, serrate or crenate, but the teeth are not distinctly pronounced, the adjective Obscurely is prefixed. For this word that of "Obsoletely" is prefixed when the leaf possesses a relationship such as to make it probable that its ancestral forms were more strongly characterized by this condition (Fig. 448). When the teeth and their sinuses are all connected in such a way that the margin represents a wavy line, the latter is called Repand or Undulate, or Sinuate (Fig. 450). When a leaf is so deeply toothed that the sinuses reach well toward the middle portion (Figs. 451, 452, 453) the term Lobed is substituted for those above defined. When the division, by a sharp sinus, extends more than half way to the middle, yet not very near to the midrib, it is called Cleft (Fig. 454A). When reaching almost to the midrib (Fig. 454) or to the base in case of a digitate leaf (Fig. 455), it is called Parted, and when all the way, Divided (Figs. 456 and 460). The divided leaf is, however, not the same as the compound leaf, inasmuch as the separation of its blade into distinct leaflets is not complete, as our preceding description calls for. The cleft, parted or divided leaf, is either Pinnatifid (Figs. 454, 463a, &c.) or Palmatifid (Figs. 455 and 456), according to the character of its venation. In all forms of lobed, cleft, parted or divided leaves, it is necessary that the detailed characters of the lobes and of the sinuses should be specified. The lobe may be acute, while the sinus is rounded (Fig. 453), or the reverse may be true (Fig. 451), or both may be acute or both obtuse (Fig. 452). The sinuses as well as the lobes frequently possess definite and characteristic outlines, indicated by terms such as have already been defined in connection with the leaf. When the teeth and sinuses are outlined by straight lines and sharp terminations, as though notched out by a pair of scissors, the margin is said to be Incised (Fig. 454a). When the divisions and sinuses are long and narrow in addition to being incised, it is called Laciniate (Fig. 456). When the margin of a leaf is turned downward or backward or rolled backward, it is said to be Revolute. Ordinarily the revolution is very slight (Fig. 457), but occasionally, particularly upon drying, it will be found extreme, each half of the leaf forming a roll, the two meeting back of the midrib (Fig. 458).

Before proceeding to speak of the forms of compound leaves, it should be stated that when one of the terms above defined

(and the same is generally true of descriptive terms used in other parts of the work) terminates in the ending *ate* or *oid*, it sometimes indicates that the condition tends toward but does not quite reach that named by the term to which the ending is appended. For example, triangulate

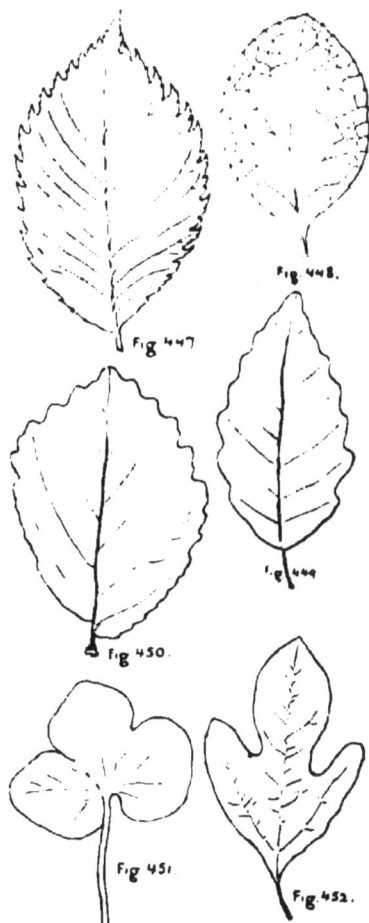

Fig 448.

Fig 447.

Fig 449.

Fig 450.

Fig 451.

Fig. 452.

means inclining toward triangular. The student will also note that between nearly all the forms of leaves and the characters indicated by the terms above defined, there are intermediate forms connecting them with others. Inasmuch as it is necessary in description for such forms to be indicated, the method is resorted to of

Fig. 453.

Fig. 454.

Fig. 454 A.

Fig. 455.

Fig. 456.

Fig. 457.

Fig. 457.

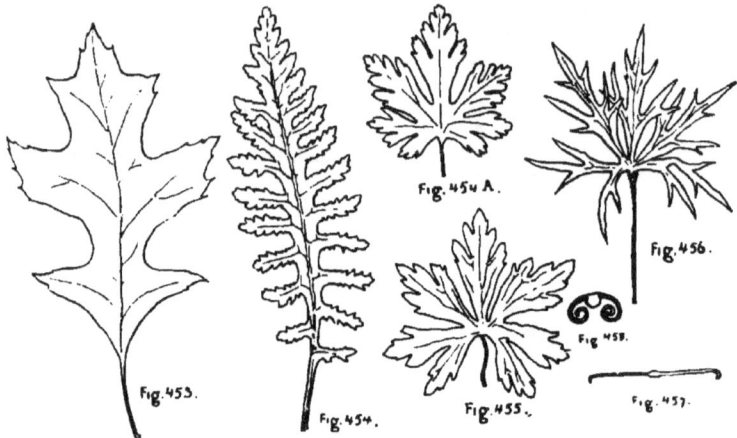

employing the two terms connected by a hyphen. Thus, Lance-ovate, or Ovate-lanceolate (Fig. 412a) indicates that the form is intermediate between lanceolate and ovate; crenate-dentate and serrate-dentate are similarly employed. A similar intermediate condition is sometimes indicated by prefixing the term *sub*, thus sub-cordate, sub-sessile, sub-acute. Other intermediate terms very commonly employed are acutish and obtusish.

A continuation or extension of the pinnately or palmately parted condition into that of the true compound leaf gives us the Pinnate (Figs. 391, 465 and 466), or, on the other hand, the Palmate or Digitate leaf (Fig. 459). Before proceeding to define the distinct forms of the two classes, we note that it is not always possible to identify them with readiness. For example, the ancestral form of the leaf of the orange was pinnate, but at the present time we find that only the terminal leaflet remains, there being usually at the base more or less of an indication of the two lateral leaflets which once existed (Fig. 460). Such a leaf cannot, therefore, be properly designated as simple, and we designate it as a Unifoliolate compound leaf. Compound leaves with three leaflets, usually designated as Trifoliolate, frequently give us considerable difficulty in determining whether they are pinnately or palmately compound. The question is to be decided in accordance with the point at which disarticulation occurs. If palmate, the base of the blade must be

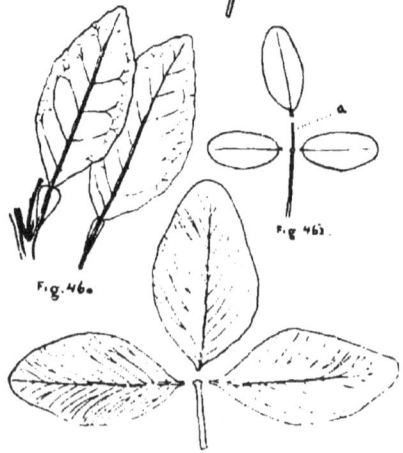

Fig. 459.

Fig. 461.

Fig. 460.

the point at which the three petioles separate, so that when disarticulation occurs no rhachis will remain extending beyond

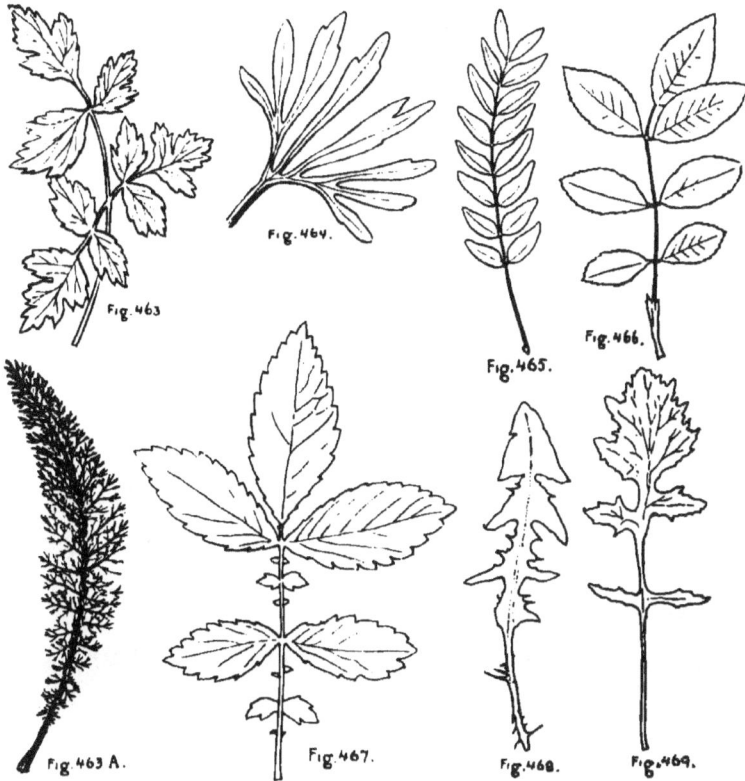

Fig. 464.

Fig. 463

Fig. 466.

Fig. 465.

Fig. 463 A.

Fig. 467.

Fig. 468.

Fig. 469.

the point of attachment of the two lateral leaflets (Fig. 461). In the pinnate form such a rhachis (Fig. 462a), although frequently very short, does exist. In the natural order, Leguminosae, the question of whether a leaf is pinnately or palmately trifoliolate is of fundamental importance in classification. A three-parted palmately compound or divided leaf is called Ternate; a five-parted one Quinate, a seven-parted one Septate. A palmatifid (or palmate) leaf, with very narrow divisions, is called Pedate (Fig. 464). If the divisions of such a leaf are similarly compound or divided, appropriate terms are formed, such as Bi-ternate (Fig. 463), Tri-ternate, and so on. Similarly named subdivisions of the pinnate form exist, the bi-pinnate (Fig. 463a), tri-pinnate, and so on. The primary leaflets of a pinnate leaf are called Pinnae, secondary ones and

those of higher rank, Pinnules. These terms are also sometimes applied to the similar divisions of pinnatifid leaves. Just as we have found that the number of pairs of primary veins of the simple leaf is generally characteristic of the species, so we find that the number of pairs of pinnae, technically known as Jugae, is equally so. This number, therefore, should always be stated, the leaf being designated as Bi-jugate, Tri-jugate, Multi-jugate and so on. Two classes of pinnate leaves are recognized, in accordance with their termination in a pair or a single terminal leaflet. Those ending in a pair (Fig. 465) are called Pari-pinnate, Even-pinnate, or Equally-pinnate, the others (Fig. 466) Impari-pinnate, Odd-pinnate or Un-equally-pinnate. When the divisions of a pinnate or a pinnatifid leaf are alternately large and very small (Fig. 467), it is called

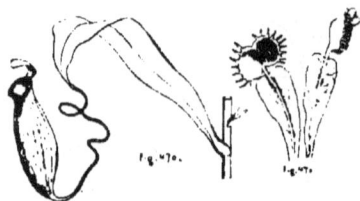

Interruptedly-pinnate or Pinnatifid. When the leaflets or divisions are turned backward so that they point more or less in the direction of the base (Fig. 468), the leaf is Runcinate. When the terminal division is very much larger, especially broader, than the lateral, the leaf is Lyrate (Fig. 469).

Coming now to consider the subject of characteristic modifications in the form and function of the leaf, we note that some of them refer to the entire leaf, others to its individual parts. We also note that in some of the modifications the entire leaf or one of its parts retains the ordinary functions of, absorption and assimilation, the new function being added thereto either by partial change of the entire leaf, or the complete modification of one or more of its parts, while at other times the original functions are entirely suppressed.

The function of absorbing and assimilating the ordinary forms of nutriment is sometimes supplemented by that of absorbing and assimilating animal tissue. In this case the leaf provides special forms of apparatus for enticing, intoxicating or mechanically catching, killing and digesting the animal, commonly an insect. One of these forms is illustrated in the pitcher plant (Fig. 470), in which one portion of the leaf becomes converted into a vessel containing liquid of variable origin and complex composition. Upon the outer portion of the pitcher a line of glandular tissue stretches downward. The insect feeds upward along this line of secretion, which so changes its nature toward the apex of the pitcher, that at the time that the insect reaches that point he is more or less intoxicated, and on crossing the margin, or quickly thereafter, falls into the liquid and is drowned, digestion promptly occurring by means of enzymes excreted into the liquid by special glands located upon the inner face of the pitcher.

Another form is the well known Venus' fly trap (Fig. 471), which secretes a nectar by certain glands which surround its margin. The insect alighting upon this point is instantly seized through the spasmodic coming together of the two lateral halves of the leaf, which act precisely like the jaws of a trap. Thus secured, digestive fluids are immediately poured forth from special glandular tissues on the leaf surface and digestion and absorption take place. That the nutrients thus absorbed are of service to the plant has been proven by elaborate experiments, in which the effects of such feeding have been esti-

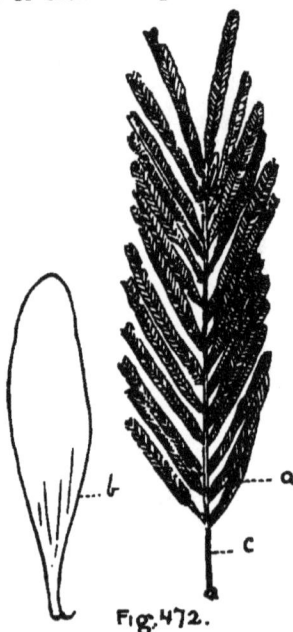

Fig. 472.

mated and compared with the reproduction by other similar plants, similarly treated in all respects except that they were deprived of this form of food.

In other cases the plant being nourished by means of fully prepared nutrients absorbed from other leafy plants (host-plants) upon which they are parasitic, the leaves lose the chlorophyll tissue upon which their ordinary functions depend and are known as Etiolated leaves. They become reduced in size and scale-like in form.

Plants which grow in excessively dry or desert regions, and which are thus very liable to suffer from excessive evaporation, ordinarily have their leaves modified in some way so as to guard against this tendency. They may become merely reduced in size or may be otherwise modified so as to reduce the amount or the degree of activity of their epidermal tissue, or they may disappear altogether, or become transformed into organs of a different character. In one of these forms the leaf becomes converted into a spine, or a group of spines, each consisting of one of the teeth. In this condition the leaf serves an important function in protecting the plant against destruction by desert animals. At other times the blade (Fig. 472a) entirely disappears, a new blade (Phyllodium, Fig. 427b) of much less activity as an evaporating organ, becoming formed by the flattening out or expansion of the petiole (c). A phyllodium is readily distinguished from a leaf blade in that its broad surfaces are directed laterally instead of vertically, as in the true lamina.

Leaves or their petioles frequently become modified into floating organs in aquatic plants, as in the case of the blad-dery-inflated petioles of the Eichornia (Fig. 473a).

Somewhat similar inflated organs exist upon the petioles of some plants and serve as the homes of colonies of ants, which are efficient in protecting the plant against the attacks of certain animals (Fig. 474a).

The office of climbing is frequently performed by a portion of the leaf. In some cases, as the Clematis (Fig. 475), the petiole of the leaf becomes twining for this purpose. At other times the apex of the rhachis (Fig. 476) becomes a tendril, either simple or branching, while at others the entire leaf becomes thus modified. In the Smilax (Fig. 477) it is the stipule which is thus changed. In other cases (Fig. 478) climbing is effected by means of hooks developed upon some portion of the leaf.

Besides protecting the plant by becoming converted into spines or spine-bearing organs, as above described, the leaf is subject to various other modifications having this object in view. Reference has already been made to such modifications in the form of bud scales. For the protection of the flower exists the epicalyx and such scales, called Floral Leaves or Bracts, as have been described in our opening account of the flowers of the willow.

Floral leaves or bracts do not always exist merely for purposes of protection. In very many cases they are functionally a part of the flower structure, surrounding either single flowers or clusters of flowers, and serving by their large size or brilliant colors, or both, to attract insect-visits, precisely the same as has been described in reference to the perigone. Through the floral bracts thus modified we get a direct transformation into the parts of the perigone as has already been sufficiently explained. It is also important to note that a direct relation is to be traced between the definite arrangements of foliage and floral leaves, as will be considered under Phyllotaxy, and the arrangement of the parts of the flower itself. So the characteristics of praefloration are seen to be directly dependent upon the phyllotaxy.

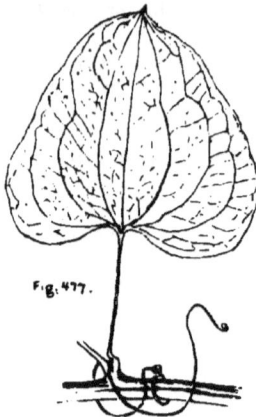

Fig. 477.

Fig. 478.

PHYLLOTAXY.

In view of the established fact that the development of the branches follows that of the leaves, it becomes clear that the arrangement of the latter determines the entire symmetry of the plant, with all the far reaching consequences in connection with both vegetation and reproduction. Certain definite laws of phyllotaxy having been ascertained, the forms resulting become, in their different manifestations, of nearly fundamental importance in classification and in diagnosis. We find that either one or more than one leaf is developed from a node. In the latter case the arrangement is called Verticillate or Whorled, and the circle a Whorl or Verticel. If the Whorl contain but two members, they are called Opposite — that is, the centres of their points of insertion are separated by one-half the circumference, or their Divergence is 180 degrees. Usually the other nodes are similarly clothed, except that in all of the higher plants the leaves of each pair Decussate with those of each adjacent pair—that is, a leaf of one whorl is over the centre of the sinus of that next below and that next above (Fig. 479). Four vertical rows (Orthostachies) of leaves thus appear upon such a stem (Fig. 480). If, instead, there be three leaves to the whorl, six orthostachies will result; if four, eight; and so on. It frequently happens that the number of leaves in the upper or lower whorls will contain only half the number of leaves in the others,

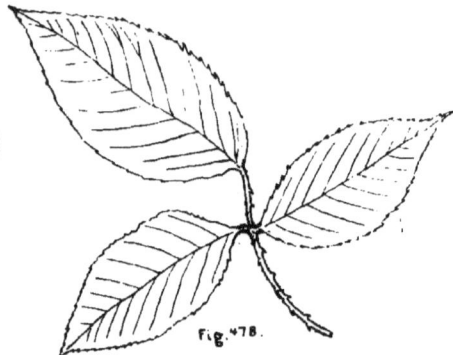

and still higher up the whorled arrangement may be lost, the leaves becoming arranged as in the form next considered.

By the other arrangement the nodes produce solitary leaves, so that each leaf is successively produced at a higher level. If a line be traced from the point of origin of one leaf to that of the one next above, and continued in the same direction, so that it exactly meets the point of insertion of another, and then of another, and so on, it will at length meet one exactly over the point of starting—that is, a second leaf in the same Orthostachy (Fig. 481). It will then be found that the line followed is a spiral, which has passed once or more around the stem. Such a spiral is called a Cycle, and if its line be continued it will form other similar cycles above and below. It is observed that a cycle will be limited by two adjacent leaves of one Orthostachy. Thus, if leaf No. 4 is the next in the orthostachy, to which leaf No. 1 belongs (Fig. 482), three leaves will belong to that cycle. A cycle containing three leaves makes but one turn of the stem. A cycle is expressed in the form of a fraction, its numerator indicating the number of times it encircles the stem, its denominator the number of leaves which it includes, so that the cycle last described must be indicated by the fraction one-third. If the next leaf in the same orthostachy as No. 1 be No. 6 (Fig. 483), then that cycle will contain five leaves. A cycle containing five leaves makes two

Fig 479

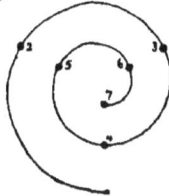

Fig. 482

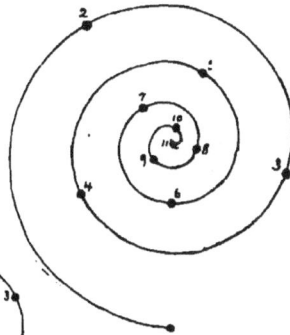

Fig. 483.

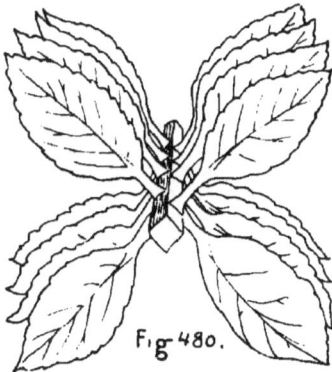

Fig. 484.

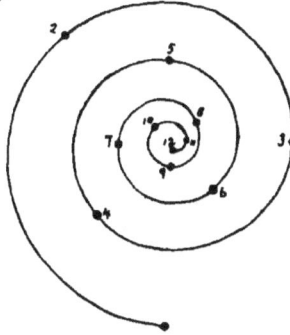

Fig 480.

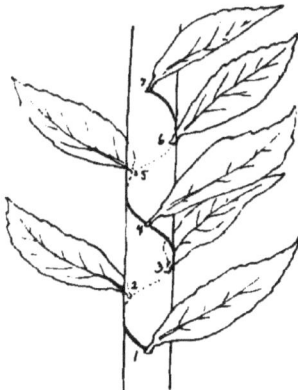

Fig. 481.

circuits of the stem, so that its exponent will be two-fifths. If the second leaf were No. 9 the appropriate fraction would be three-eighths, the cycle making three turns and containing eight leaves (Fig. 484). It will thus be observed that these fractions form a series, in which each possesses a numerator equal to the sum of the numerators of the two preceding and a denominator equal to the sum of the denominators of the two preceding. No cycles occur among the higher plants with which we are concerned, which can be indicated by any fraction not thus formed.

Noticing these fractions still further, we observe that the denominators will indicate the number of orthostachies upon the stems which they represent, and that the value of the fraction will represent the divergence of, or part of a circle between, any two leaves adjacent in the cycle or spiral—that is, the number of degrees between such leaves will equal that fractional part of 360 degrees.

As to the direction which the spiral takes, it may be either from right to left or from left to right. It is supposed that each kind of plant, at least of the higher classes, produces two forms or "castes,"

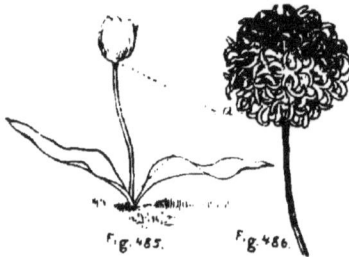

Fig. 485.   Fig. 486.

depending in some not yet perfectly determined way upon the relative positions of the respective ovules from which they originate. The tendency of these two castes to manifest their growth or development in opposite directions has been called Antidromy.* Among numerous other phenomena attributed to antidromy is this starting of the leaf-spiral in opposite directions in plants of the two castes of any species with this form of phyllotaxy.

Occasionally leaves appear to be irregularly disposed upon the stem—that is,

tered, and the explanation is different in different cases. When a stem is so shortened that the leaves are crowded upon it in the form of a regular rosette, as in the house leek, the arrangement is called Tufted. When similarly short, but the leaves few and irregularly crowded in a little bunch, the arrangement is Fascicled.

The two regular forms of leaf arrangement above described can be traced in greater or less perfection through floral bracts and involucres and into, and in many cases partly or wholly through, the flower itself. While such arrangement in the flower is in many cases entirely verticillate, and in most cases partly so, it has been quite clearly shown that many flowers have certain of their parts arranged upon the spiral plan.

ANTHOTAXY.

That part of a stem or branch which bears the flowers, or the flower when solitary, is more or less distinctly modified in form, surface and extent and character of branching, and frequently also in the direction taken in the arrangement of its parts. In connection with its flowers it is called the Infloresence. The portion of an inflorescence which is below its lowest point of branching or flowering, or below the flower when solitary, is called the Peduncle (Figs. 485 and

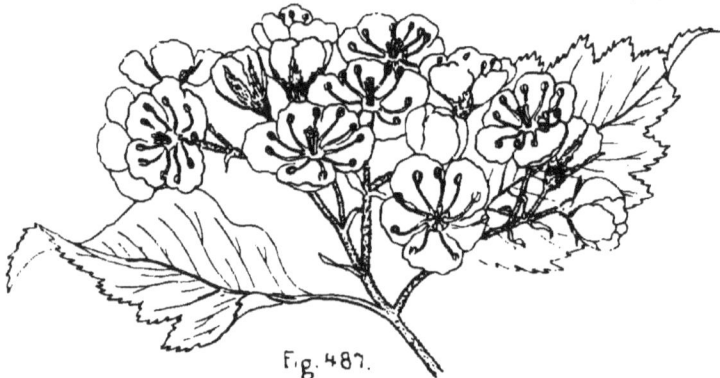

Fig. 487.

they are not whorled, nor does the law of alternate phyllotaxy appear to apply to them. This arrangement is called Scat-

*See article by Professor George Macloskie, in Bull. Torr. Bot. Club, XXII., 379.

489a). This name is also applied to the corresponding portion of a branch of an inflorescence if that branch bear more than one flower, it being in that case a Secondary Peduncle (Fig. 494d). If the

Fig. 488.

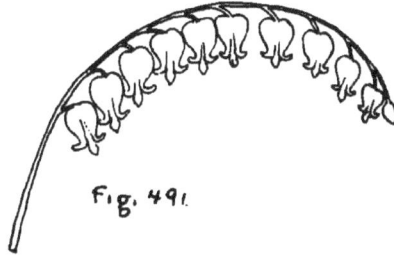

Fig. 491.

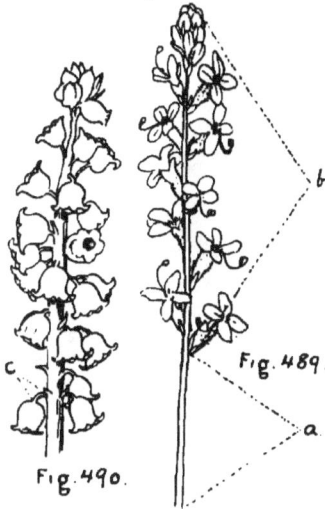

Fig. 489.

Fig. 490.

peduncle is continued above its first point of branching, in the form of a central support along which the succeeding branches are arranged, this portion is called the Rhachis (Figs. 489b and 492a). A peduncle which rises directly from or near the ground is called a Scape (Fig. 485a). The stem of one of the individual flowers of an inflorescence of more than one flower is called a Pedicel (Figs. 490 and 494c). A flower or an inflorescence may be devoid of pedicel or peduncle, when it is Sessile. The arrangement of flowers is called their Anthotaxy, and this name is also applied to the study of inflorescences.

The arrangement of the inflorescence-leaves and their floral branches, while based upon the phyllotaxy, and traceable thereto in most cases, exhibits more or less real or apparent departure therefrom,

and calls for special designations and classification. The forms are divided into two series in accordance with the apical or lateral location of the initial flower—that is, the flower which is first in order of development. If the terminal bud develop into a flower (Fig. 485) its further extension is impossible, except by the rare and abnormal process of Proliferation. Inflorescences so limited are therefore called Determinate or Definite. Although vertical extension of the original stem of a determinate inflorescence is not possible, it can apparently take place through the branches, the same as in other sympodia. The effects of such development are the same as in other forms of sympodial growth in which there is a transformation of the apex of the original stem—as, for instance, in our explanation of such a mode of development of the tendril (Fig. 356). To apply this principle in the case of an inflorescence we have only to assume a flower developed at the tip of every branch in Figs. 373 A, B and C. Flower A would develop first; B, although the second in order, and hence a branch, and afterward C, would be more elevated, and would thus seem to prolong the vertical extension of the stem. The development, being successively by nodes whose original points of origin were successively lower than that of the terminal flower, is structurally and really Descending or Basipetal, even though by the upward growth of the successive branches they be at successively higher levels, the order apparently in the opposite direction. By the development at each node of a pair of opposite branches we get the apparent bifurcating or dichotomous form (Fig. 373A). If but one branch grow from a node, and these successively from right to left, the zig-zag or Flexuose form of rhachis is produced (Fig. 373C), and if constantly from the same side, or apparently so, the Circinate (Fig. 373B). The descending or basipetal nature of the definite inflorescence is clearly shown when the successive branches remain short, each successively developed flower remaining at a lower level than that which preceded it. Instead, however, of assuming either of these two states, in which the flowers remain at different levels, the branches may radiate and elongate to different degrees, ceasing their elongation when their flowers have been brought to a uniform height, so that a more or less flat-topped inflorescence results, the order of development being form the centre outward, or Centrifugal, as in the branches of Fig. 494. This form represents the true Cyme, and because of their relationship to it this entire series of inflorescence is often denominated the Cymose. It will thus be seen that in different forms of the cymose inflorescence we may have the flowers (a) all brought at length to a uniform level (b), those successively later developed brought to successively higher points, and (c) those left at successively lower levels. This fact demonstrates that the cymose or descending nature of an inflorescence cannot be determined by noting the relative heights of the flowers themselves, but only by noting the order of their development.

In the other series the first flower to develop is structurally the lowest of the cluster, the succession being upward, Ascending or Acropetal (Figs. 489 and 490). If the successive branches develop less rapidly than their predecessors the result is again a flat topped inflorescence with the development from the outside to centre, or Centripetal (Figs. 487 and 488). The branches and flowers may be separated on obvious peduncles and pedicels, or these may be not apparent, the flowers being sessile. In accordance with the characters above explained, we obtain the following simple forms of anthotaxy:—

Series 1.

Ascending, Acropetal, Indefinite, Indeterminate, Centripetal or Botryose Forms.

A. With the rhachis not elongated.
   1. The Capitulum or Head, with the flowers, and branches if any, sessile or so regarded (Fig. 486).

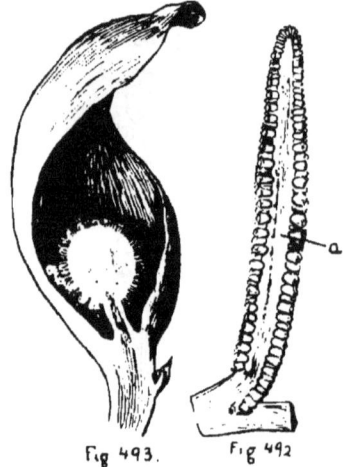

Fig 493.   Fig 492

Fig. 494

   2. The Corymb, with the rhachis manifest, though short, and its pedicels or branches elongated so as to produce a flat-topped inflorescence (Fig. 487).
   3. The Umbel, similar to the Corymb but with the rhachis not manifest, so that the pedicels or branches all appear to

start from one point at the summit of the peduncle (Fig. 488).

B. With the rhachis elongated.

4. The Spike, with the flowers, or branches, if any, sessile or so regarded (Fig. 489).

5. The Catkin or Ament, a spike, with slender rhachis and bearing usually staminate or pistillate flowers, crowded and subtended by scales (Figs. 8, 12 and 15).

6. The Raceme, similar to the spike or ament, but having the flowers pedicelled (Figs. 490 and 491).

When either the head or spike possesses a thick, fleshy rhachis it is called a Spadix (Figs. 492 and 493).

Series 2.

Descending. Basipteal. Definite. Determinate. Centrifugal or Cymose Forms.

1. The Glomerule, corresponding to the head in all respects save that the central flower first develops.

2. The Fascicle, similar to the glomerule except that the flowers are few and loosely clustered.

3. The Cyme. Similar to the corymb or umbel, save that the central flower is the first to develop (Fig. 494).

4. The Scorpioid Raceme. Similar to the raceme, except that each successive node and flower upward is lateral to that next below. The apex of the scorpioid raceme is circinately coiled (Fig. 373B).

Before proceeding to consider certain special forms and modifications of the inflorescences above defined, it should be remarked that most of the forms may be compound. By this we mean that the cluster is made up of a number of branches whose order of development is the same as that of the elements of which they are composed. That is, the raceme may possess a number of branches, each of which is a smaller or secondary raceme, or if not a raceme, at least a small inflorescence of the ascending or centripetal form. Similarly an umbel may be made up of branches, each of which is a smaller umbel, the Umbellule. A cyme will be made

up of cymules, and so on. A Panicle is a compound raceme which assumes the form of a pyramid. Any form of inflorescence not · a true panicle, but assuming the shape of one, is styled Paniculate.

Complex forms of inflorescence differ from the compound in that the order of development of the several flowers upon a branch is of the opposite kind from that of the several branches themselves. For example, the Thyrsus or Thyrse is a paniculate form in which the lowest branch is the first to develop flowers, so that the order of development of the branches is ascending, but within a branch the terminal flower will be the first to develop, so that the order of development of its flowers is descending.

The term Hypanthodium has already been defined in considering the forms of the fruit, under Multiple or Collective Fruits. The same term is applied to an inflorescence yielding the collective fruit of that name (Fig. 495). It is in reality nothing more than a head closely subtended, surrounded or enclosed by an involucre (a). The hypanthodium is characteristic of the great family Compositae and is of so much importance in classification that its modifications call for special attention. The involucre should be studied as to whether it is single, double or multiple—that is, whether it consists of one, two or more circles of scales; as to whether these are equal in length or whether the outer or inner are successively shorter; whether they are entirely free and distinct, or adnate by their bases or connate by their margins; as to whether they are appressed, or with more or less of their apical portions recurved or spreading; especially as to the general form of the involucre as a whole, the terms used being the same as those previously applied to the perigone, and as to the characters of the individual scales, these being practically the same as those which have already been considered in connection with the leaves. The body consisting of the combined tori of all the flowers of the hypanthodium is called a receptacle (b). It is to be studied as to its being solid or hollow; as to its general form, and especially the form of its upper surface, whether concave, plane, convex,

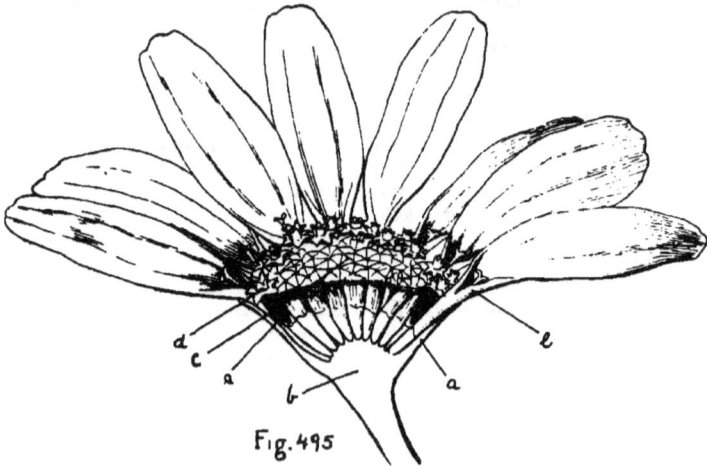

Fig. 495

rounded or conical; as to its being smooth in surface, honeycombed or otherwise pitted (foveolate), and if the latter the special characters of the pits and their margins; and as to its being naked or clothed with hairs or scales, and the characters of the latter. The head is then to be considered as to the character of its flowers. If these are all similar, the head is said to be Homogamous; if different, Heterogamous. If the flowers are all ligulate, the head is Liguliflorate. If it possesses a disk (c), of tubular flowers (d), it it Discoid. If this disk is surrounded by one or more circles of ligulate flowers called Rays (e), it is Radiate. If the ray flowers and disk flowers are of the same color, the head is Homochromous; if different, Heterochromous. The flowers must next be studied as to their sex. The ray flowers are commonly pistillate, while the disk flowers are perfect, or the disk flowers may vary among themselves in this particular. Very commonly the ray flowers are entirely neutral. Even if pistillate they may be sterile. If both classes of flowers are fertile, the akenes which they produce may be heteromorphous, those of the disk being commonly compressed, those of the rays commonly triquetrous. Occasionally the heads are dioecious or monoecious. In one tribe of the Compositae the flowers are bilabiate. The character of the pappus (Figs. 61 to 68B) is invariably of the utmost importance, as are the forms of the style-branches and the appendages borne by these at the apex and by the anthers at apex and at base (see Androecium and Gynoecium). Many special terms are applied to the forms of inflorescence leaves or bracts. A single large bract subtending, surrounding or enclosing a spadix, is called a Spathe (Fig. 493).

One group of natural orders, the grasses and grass-like plants do not possess any obvious perigoinne, its place being supplied by peculiarly formed, adapted and arranged bracts in the form of scales or chaff, and technically called Glumes, which give to this group of orders the title Glumaceae. In the rushes these glumas really are a true perigone, which is trimerous. In the sedges (natural order Cyperaceae, Fig. 496), the scales (a) are solitary, subtending each flower. In the grasses (natural order Gramineae) the glumes are arranged in pairs, each pair subtending a short branch, which may bear only one, several or many flowers, the whole known as a Spikelet (Fig. 497). Typically there is besides the two glumes of the spikelet (a) an additional pair of scales for each flower. Thus, if there be but one flower in a spikelet, it possesses two pairs of scales. If more than one, then there is a separate pair of scales for each flower, besides the one pair pertaining to

Fig. 496.   Fig. 497.

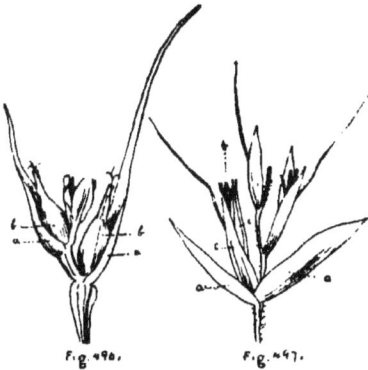

the spikelet as a whole. The scales of the spikelet are called the Glumes, Glumes Proper, or Lower Glumes; those of the individual flowers (c) Palets or Upper Glumes. Much complexity in the relations of the glumes ensues as a result of suppression of both glumes or both palets, of one of either or of each, or of two of one and one of the other, and so on. The character of the individual glumes must be carefully studied, as in the case of the involucral scales of the hypanthodium. The character of the terminal appendages which they bear is of special importance.

With this study of the inflorescence we are brought again to the individual flower, with the study of which we commenced.

AN OUTLINE

OF

PRACTICAL PLANT ANATOMY

BY

SMITH ELY JELLIFFE, M. D.

Professor of Pharmacognosy and Director of the Microscopical Laboratory of
the New York College of Pharmacy.

INTRODUCTION.

THE following outline of plant anatomy is intended more particularly for the use of students of medicine and of pharmacy, and, as such, is offered in the hope that it may be of service to beginners in the study of plant tissues as applied to their practical determination in the official drugs.

My thanks are due to Prof. A. Tschirch, of Bern, for his generous consent to use cuts from his works, and also to Prof. Coblentz and Dr. Schneider and their publishers for the use of cuts.

SMITH ELY JELLIFFE, M. D.
231 W. 71st St., N. Y. City.

TABLE OF CONTENTS.

INTRODUCTION 106

CHAPTER 1—The Microscope. Its parts and principles.... 107

               Technique of plant histology...... 114

CHAPTER 2—The plant cell in general 115

               The cell contents........................... 115

               The cell wall 115

CHAPTER 3—Cell contents.   Classification.................... 115

               (a) Unformed nitrogenous contents.. 116

               (b) Formed nitrogenous contents ... 116

               (c) Non-nitrogenous contents....... 118

               (d) Cell sap........................ 121

CHAPTER 4—The cell wall and its modifications, morphological and chemical.................................... 123

CHAPTER 5—The tissues in general.   Classification.......... ... 124

               (I)   Formative tissues........ 125

               (II)   Protective tissues.... 125

               (III) Tissues of nutrition 125

               (IV) Reproductive tissues...... 125

CHAPTER 6—The tissues. Formative tissues 126

CHAPTER 7—The tissues. Protective tissues 127

CHAPTER 8—The tissues. Tissues of nutrition................. 134

CHAPTER 9—The tissues. Reproductive tissues 147

CHAPTER 10—Micro-chemical reactions........................ 147

OUTLINE

OF

PRACTICAL PLANT ANATOMY

CHAPTER I.

THE MICROSCOPE.

While it is undoubtedly true that one must look to the pharmaceutical chemist for both the quantitative and qualitative determination of plant constituents, yet the skillful microscopist can often shorten the path in analysis and keep a check upon the chemical results. Many plant constituents are directly recognizable by means of the compound microscope, and when micro-chemical tests are employed in addition there are few plant products that cannot be detected, even with more certainty than by chemical analysis. The details of such tests form the matter of many standard

microscope, its parts, both mechanical and optical, and its principles, with additions of a technical nature to enable the student to apply the instrument in practical drug examination.

Microscopes are of two kinds, the Simple and the Compound.

The Simple Microscope, or magnifying glass, consists of one or several double convex lenses, and gives a direct image of the object. This image is erect; the field of view is generally extensive, and the magnification is limited, varying from one to twenty times the size of the object.

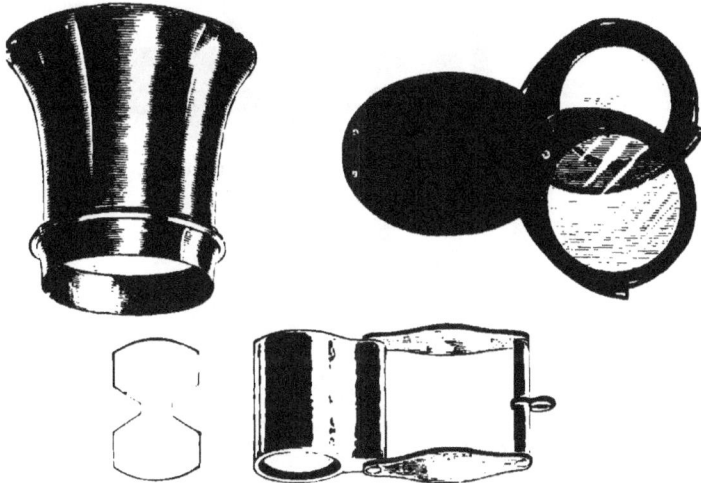

Fig. 1. Types of Simple Microscopes, or Magnifying Glasses.

works of reference, which cannot be consulted without a working knowledge of the microscope. The object of the present chapter, therefore, is to describe the

The Simple Microscope is invaluable in investigations of plant structures, especially of the grosser parts. In the determination of leaves and flowers the

Eye-Piece or Ocular.

Draw-tube.

Collar

Coarse Adjustment.

Fine Adjustment.

Nose-Piece.

Arm

Pillar.

Clips.

Stage.

Sub-stage.

Diaphragm.

Pillar.

Mirror.

Base

Abbe Condenser.

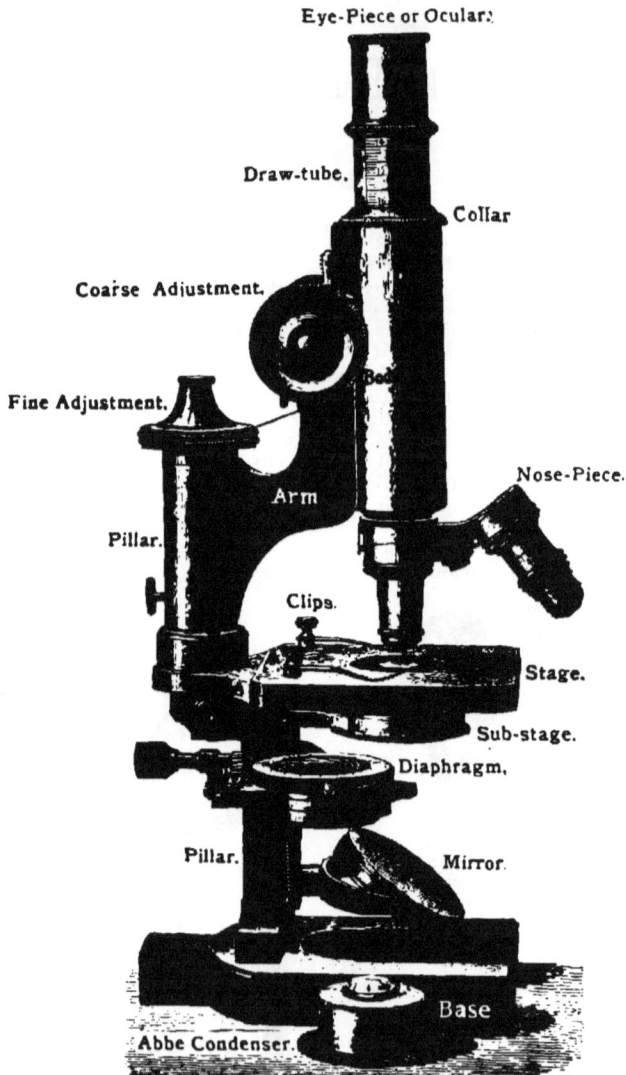

Fig. 2. Compound Microscope.

pharmacognocist cannot do without it.

The Compound Microscope is an optical instrument constructed for the purpose of magnifying small objects, and differs from the Simple Microscope in that it magnifies the magnification of the object. Compound Microscopes may be very simple in construction, consisting of but few lenses, but the principles involved are always the same.

In the Compound Microscope the image is inverted, the field of view is small, the illumination or light coming from the object is diminished, and the magnification is much increased.

The parts of a Compound Microscope may be conveniently divided into two groups, the mechanical and the optical. A reference to the accompanying illustration will serve to locate the parts named and described.

The mechanical parts are the base, or foot; the pillar, below stage and above stage, with a hinge joint between; tube, or body; draw tube; rack and pinion, or coarse adjustment; micrometer screw, or fine adjustment; stage; nose-piece; stage clips; diaphragm; mirror bar; sub-stage, if present.

The optical parts are oculars, or eye-pieces; objectives; mirror; condenser.

The Mechanical parts or parts of the Stand proper.

The Base or Foot is that part which supports the microscope. It is made in various shapes, triangular, circular or horse-shoe. It should be firm, and heavy enough to allow the microscope to be tilted upon its axis and still stand steady.

It bears the Pillar, which has a portion below and above the stage, which is broken by a hinge joint (not always present), which allows the microscope to be inclined at any angle.

To the pillar is attached the Stage and the Arm. The Stage is generally round or square, and should be firm, flat, and large, at least four inches across. It is perforated at the centre by an aperture, which allows the light to pass through it to illuminate the object. It also has a pair of Clips to hold the glass slide in place.

The Arm bears the Body or Tube. A firm arm is an essential to a good microscope.

The Body or Tube is attached to the arm by means of a Rack and Pinion piece. This is called the Coarse Adjustment, and permits the tube to be lowered or raised. The Fine Adjustment consists of a micrometer screw situated at the top of the pillar.

The Tube frequently has a draw tube, into the upper end of which the Ocular or Eye-piece is set. To the lower end of the tube the Objectives are attached.

In most modern instruments there is added an especial apparatus called the Nose-piece, to which two or more objectives may be attached, any one of which may thus be brought into position.

Below the stage there may be Sub-Stage, which may be provided with a Condenser, the Abbé being the best. From the lower portion of the stage near the pillar the Mirror Bar descends, car-

Fig. 3. Iris Diaphragm.

rying the Mirror, and in some instances a Diaphragm. In some instruments the diaphragm is situated just below the aperture, and may be of several types; the most useful one is the Iris Diaphragm, as displayed in the cut.

In the best of modern instruments the whole sub-stage arrangements are bound together in one mechanism, as in Fig. 4.

The Optical Parts:

The Ocular or Eye-Piece is set into the upper portion of the tube. There are two types of eye-pieces, the negative, or Huyghenian, which is the more commonly used, and the positive, or Ramsden, now rarely seen in microscopic use. In the more advanced types of instruments there is the so-called compensating ocular, which is, however, only used with the newer apochromatic objectives.

The action of the eye-piece is that of a simple magnifier, but it magnifies the real image which the objective produces, as if that image were the original object.

Oculars are designated as high and low, and are measured in inches or millimeters; the longer the ocular, the lower or weaker it is in magnifying power, and vice-versa. (Fig. 5.)

The Objective is often called the most important part of the instrument, which is, in many senses, true, for the great advances that have come about in the biological sciences have been mainly due

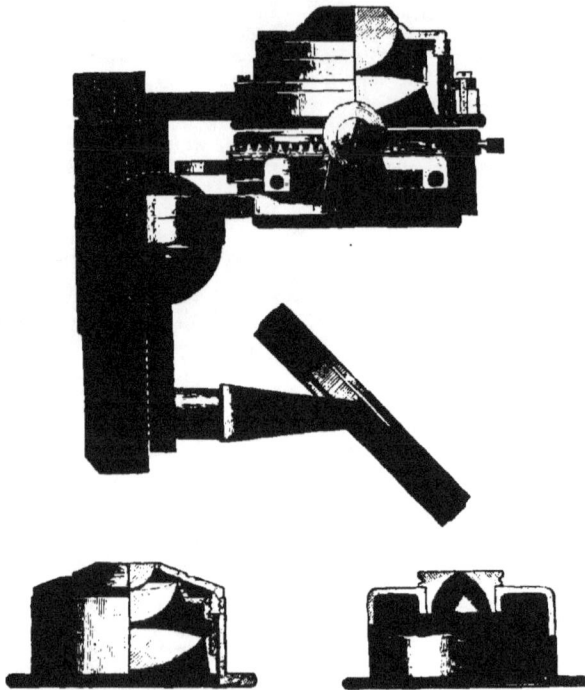

Fig. 4. Sub-Stage Condenser.

to the advances that have been made in the construction of the objectives.

The objective is placed at the lower end of the tube, either fitting directly into the tube, or into the nose-piece by means of what is known as the Society Screw, which is an internationally adopted size, so that any objective can be used in any stand.

The objective is made up of a series of lenses varying from two to four, and works in a combination as a simple magnifier; it forms a real inverted image in the tube, which is further magnified by the ocular.

Objectives are designated either arbitrarily, as a, b, c, d, e, etc., or according to a system of measurements expressed in fractions in the English or in the metric system. These measurements signify that the magnifying power of any lens, say a ½-inch objective, is equal to the magnifying power of a simple lens whose focal distance equals ½ inch. In general, therefore, the lower the fraction which designates the lens, the higher is the power of magnification. For general purposes, "the smaller the front lens, the higher the power," is a good guide for the student to enable him to know which is the high power and which the low. It should also be remembered that, as a rule, the higher the magnification, the less working distance, the less field, and the less illumination. (Fig. 6.)

The Mirror is attached to the mirror bar beneath the stage, and is freely movable. It consists generally of two surfaces, a plane and a concave, the latter of which is generally used; it should be of ample width, at least two inches.

The Condenser, when present, is situated just below the stage, and if of the Abbé type, is well represented in the illustration; it is used to give a greater supply of light to the objective, and is invaluable in working with stained specimens, which are to be differentiated by means of color rather than by outline. For the best work it is an essential adjunct to the microscope. (Fig. 4.)

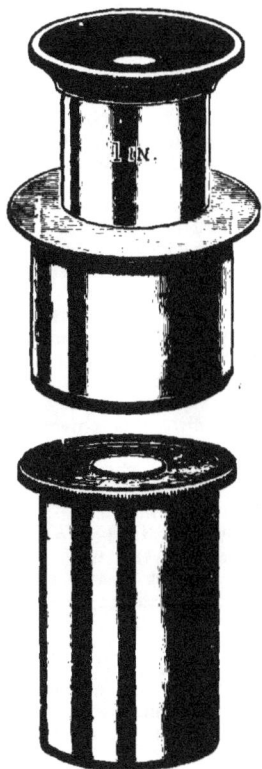

Fig. 5. Eye Pieces.

In choosing a microscope for use in the laboratory a few cardinal principles may be borne in mind. As many differences of opinion will be found among working microscopists, the question must be left to optical principles; for any microscopist, after working many years with a particular type of instrument, may become so proficient in its use that, notwithstanding radical errors in its construction, good work can be done. Individual bias should not, therefore, govern the purchaser.

Firmness and solidity are virtues in a microscope, and as these qualities are given by the base, the pillar, and the arm, these parts should be strong and

solid, not necessarily massive. We prefer the so-called Continental type, as giving compactness, firmness and solidity.

The stage should be ample; a square is perhaps preferable, and it should be thick and firm. The investigator will often use corrosive substances in his work, and the stage would better be of metal.

A coarse adjustment, preferably a rack and pinion, that will not wobble, and that will not allow the tube to descend of its own weight, is a requisite. A fine adjustment is an essential; it should work evenly and smoothly, and is preferably placed at the top of the pillar.

A nose piece, double or triple, is now considered a necessity. Care should be taken that it centres the objectives and that its joints do not leak.

As the objective is the most important part, a few words regarding its choice may be of benefit. Two errors of construction should be borne in mind; these are chromatic and spherical aberration.

Chromatic aberration is a defect due to the double action of a lens, by which it acts as a magnifier and as a prism. Acting as a prism, it decomposes the

Fig. 6. Construction of Objective.

light into its elements, and if a lens shows bands of color around the edges of the object looked at it has not been totally "corrected" for chromatic aberration.

Spherical aberration is a second defect whereby all points of an image are not brought together, so that in looking at a piece of fine wire netting, for instance, the fibers in the centre are straight and distinct, while those on the outside are found to be curved and blurred.

A certain amount of spherical aberration is necessary in high power objectives, but by means of the diaphragm the outer portions can be cut out of view.

Most modern objectives are carefully corrected for both these errors, but they should be remembered by the purchaser.

Good lenses should also possess good defining power, that is, the image should be distinct, especially at its outer borders; the diaphragm has much to do with the distinctness of the image, and care should be taken to limit the amount of light when one is testing a lens for its defining power.

Flatness of field must vary with the magnifying power and angle of aperture of the lens. A flat field is one in which all the parts of the field are in focus at the same time. Regarding the angle of aperture little need here be said, save that it represents the pencil of light that the lens is able to take in and use in forming an image; and that the angle of aperture of a lens more clearly tells its powers than the tables of magnification. Wide-angled lenses that are properly corrected are to be preferred.

Good working distance is of great importance to the microscopist in the pharmacognostical laboratory, and such objectives should be chosen that have this quality, even if a certain amount of magnification should be sacrificed. It means that there should be an appreciable distance between the front lens and the object. In low-power objectives it is of little concern, but in the higher powers it becomes important, especially to the pharmacist who may desire to look at some crystals in a fluid in a watch crystal; if he has a lens with a short working distance it may be impossible, because the objective will not focus unless stuck into the liquid, and this destroys the image, whereas a long working distance lens would give the requisite ability to manipulate the object.

It is to be taken for granted that the pharmacognocist knows how to use the microscope, but the following brief directions may be of service to the beginner.

The working table should be firm and is best about 28 or 30 inches in height. The light from an illumined cloud is the best light to be used and most microscopists prefer north light. Any light can be used, but not too much. When using a lamp the light from the ground glass globe is preferable to that from the wick.

The microscope should be placed directly in front of the observer, with the pillar facing backward. To get the light the mirror, preferably the concave side, should be so turned as to throw the beam directly up through the aperture. *All work should be begun with the low-power objective.* Having obtained the illumination of the object, which may be mounted on a slide or held in a glass evaporating dish, this object is placed as near the centre of the stage as the eye can approximate, when by means of the coarse adjustment the low power is lowered as near the object as possible without touching it; then with the eye over the ocular, the tube should be gradually raised by means of the coarse adjustment, until the object is in an approximate focus; the exact focus can then be obtained by means of the fine adjustment. In general it is not necessary to turn the fine adjustment more than one revolution.

Having observed what was necessary with the low power, the portion of the object further to be investigated should be moved so as to come in the centre of the field, the observer still looking into the microscope.

The tube is then raised and the high power substituted, and the process of focusing, as already outlined, repeated. It is essential in using the high powers to exercise great care. The objective approaches the object as near as it can be placed, and then with the eye over the ocular the tube is raised very slowly by means of the coarse adjustment until the focus is reached; then the focus is completed as before by means of the fine adjustment.

To illustrate the value of the diaphragm the following experiment may be tried. A thin transparent object should be selected and brought into focus with the low power; then the diaphragm should be opened as wide as possible and later gradually closed; it will now be seen that as the diaphragm is closed the outlines of the object become more distinct, until a point is reached when the light is so much diminished that the object is no more visible. It is a lesson of experience to make the exact compromise between the amount of definition and the amount of illumination.

Accessories.

Stage micrometer.

Ocular micrometer.

Camera lucida.

In the laboratory quantitative estimations are desirable, and to ascertain the size, as well as the shape of objects studied is a necessity. To this end the microscopist makes use of instruments for measuring. The most useful is the Eye-Piece Micrometer.

This generally consists of a circle of glass, accurately ruled, which slips into the eye-piece and rests upon a diaphragm. The accompanying figure illustrates this accessory.

The eye-piece micrometer must be standardized before using, by means of a Stage Micrometer, which generally is made upon a glass slide. In the centre of the glass slide there is a finely ruled scale, usually of one millimeter divided into 100 parts.

This serves as a standard, and the eye-piece micrometer is measured in the following way: With a given eye-piece,

compared with the divisions of the stage micrometer. It is best to take the entire 50 divisions of the eye-piece micrometer and compare their length with the divisions of the stage micrometer.

If these 50 divisions are found to measure 7 divisions on the stage micrometer, then 50 lines on the eye-piece micrometer equal 7 on the stage micrometer. Each division of the stage micrometer equals $1/100$ of a millimeter, hence the 50 lines of the stage micrometer equal $7/100$ of a millimeter, and each division of the eye-piece micrometer equals $1/50$ of $7/100$, or $7/5000$, or $14/10000 =$

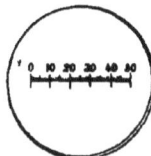

Fig. 7. Eye Piece Micrometer.

.0014 mm. = $1^{4}/_{10}$ mikrons, or micromillimeters.

Having standardized the eye-piece micrometer for both eye-pieces and both objectives, and entered a note of the values on a card for reference, it can then be used to measure any object, as, for instance, starch grains, or oil globules, or resin masses.

If the pharmacognocist is to do any research work, accurate delineations of the object seen are requisite; for this purpose some drawing apparatus will be

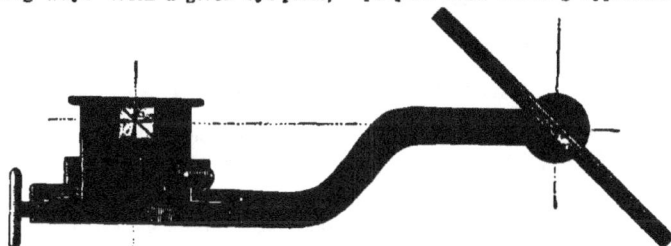

Fig. 8. A Simple Form of Abbe Camera Lucida.

say 2 inches, and a given objective, $1/6$, and a standard tube length, 160 millimeters, the stage micrometer is brought into focus. The eye-piece micrometer is then introduced and its divisions are

found of service. The well-known Abbé camera lucida is the best instrument in the market at the present time for this purpose, though other excellent instruments are to be had for a much smaller price.

The accompanying illustration represents one of the simpler forms of the Abbé cameras. The latest and most perfected instruments are to be recommended.

In drawing, the microscope must be erect and the drawing table horizontal, and the axis of the drawing should correspond to the axis of microscope, else distortion of the image will result. Special drawing tables are in the market, but are not necessities for the present purposes.

A polariscope attachment is of great service in the hands of the pharmaceutical microscopist, as it can abridge many of his investigations. It is not, however, indispensable.

The Necessary Apparatus to be used in the investigation of vegetable tissues is little.

The following articles are essential: Razor, preferably with a thin blade; scalpel, forceps, camel's hair brush, two needles, glass slides, English pattern, with ground edges are to be preferred. (These come in various sizes and thicknesses. For the beginner, a medium size is preferable, ¾-inch and No. 3 in thickness, it being less easily broken.). Glycerine, blotting paper and lens paper will be found of service. To this list of essentials many articles may be added, as section lifters and microtomes, which become necessary as soon as the student has advanced.

General Directions for Work.

The first requisite for good histological work is cleanliness. All slides, cover glasses and other apparatus in use should be kept free from dust.

Soft tissues cut differently from hard tissues, hence they require different handling. For the first a sharp razor with a thin blade should be used; for the second a sharp knife or scalpel, or a thick-bladed razor is better. If great care is taken both kinds of tissue can be cut with the narrow bladed razor, but it is wiser to be provided with both kinds and to use them appropriately. This same caution applies to the use of microtome knives.

Sections of tissues should be made with an oblique motion of the cutting edge of the razor, pulling it from heel to toe or pushing from toe to heel. Many tissues need moistening before cutting, especially the softer, pulpy kinds; for this purpose a mixture of equal parts of alcohol and water may be employed, or clean water alone. The blade of the razor may be kept moist by means of the alcohol and water mixture, although this is not always essential. The object can generally be held between the thumb and forefinger, and the blade of the razor resting upon the forefinger, it can then be drawn or pushed obliquely through the specimen. In general it is not necessary to cut sections through the entire width of the object, so that a number of small sections may be made by the oblique sawing motion of the razor before removing them to the slide for examination. Sections should be cut as thin as possible. Practice alone will enable the student to become expert in this matter. The sections are then to be removed by means of the camel's hair brush to a drop of water left on the center of the slide. Here they can be arranged by means of the brush or needles, and the cover glass placed in position. The placing of the cover glass may be done well or indifferently. After cleansing it with lens paper, it should be handled by the forceps only, and placed upon the drop of fluid containing the specimens, in as slanting a manner as possible. This avoids including air bubbles among the specimens.

After the cover glass is in position the surplus of water, if present, should be removed by means of the camel's hair brush or blotting paper; if there is not enough water under the cover glass to entirely surround the specimens, more should be added by means of the brush to the edge of the cover glass.

The slide should then be placed upon the stage, the object being as near the center as can be approximated, when the general directions for focussing should be carried out.

CHAPTER II

THE PLANT CELL IN GENERAL.

The present hypothesis of the structure of all matter is known as the atomic theory. It teaches that all matter, both organic and inorganic, is made up of particles that cannot be divided; these it calls atoms. These atoms exist only in combinations of two, three or more, when they are termed molecules. From these molecules, invisible, are built up the cells, visible, which are the microscopical units of all organisms.

The Cell is then, from this standpoint the unit of structure of animal and vegetable life; not the smallest unit, which, as we have seen, is the atom, nor the largest, which varies extensively in both the animal and vegetable worlds; but it is the smallest organized unit that has as yet been seen by the human eye aided by the microscope.

The cell may thus be taken as the unit of plant life; and with its study commences that branch of knowledge known as Vegetable Histology. Its exclusive study is termed Vegetable Cytology.

Vegetable Histology is that portion of knowledge which deals with the cells of plants, either as individuals or in their various combinations known as tissues. It classifies them and arranges them according to some logical plan of growth or of function, and inquires into their origin, their method of growth and combination, and the part they play in the economy of the plant's existence.

There are reasons to believe that in pre-historic times there was a period when the structure of cells was very simple indeed, and attempts have been made to ideally reconstruct those early forms. As time progressed and varying circumstances arose, simplicity of cause and effects gave way to complexity, and the plant world as seen to-day is one vast forest of effects, having their origin in causes remote and present.

To be rightly understood the cells of animal and plant life must be considered as an elaborate product of millions of years of manufacture. Each part has been, and is being, studied with great minuteness, and the results of the study of vegetable Cytology already fill many large volumes.

For the purposes of the present outline, the plant cell may be considered as a sac containing a large number of contents, which vary widely in physical and chemical properties.

Cell Wall. The lining membrane is called the Cell Wall. It is not always present, as in many one-celled plants, as yeast, nor is to be found in the youngest growing parts of the plant, as in the apices of stems and roots, nor in the immature pollen grains and the just fertilized cells of the ovule.

Cell Contents. While alive and growing the plant cells contain what is called the Plasma, which is a general term including a number of diverse substances to be studied under the head of cell contents. PROTOPLASM has been used extensively in this sense and as a biological conception it is proper so to use it; but protoplasm must be considered as a combination of substances rather than any one chemical entity.

CHAPTER III

CELL CONTENTS.

The following classification of cell contents has been here adopted:
1. UNFORMED NITROGENOUS CONTENTS. Including the so-called Cytoplasm of the cell body.
2. FORMED NITROGENOUS CONTENTS. Including the
    (a) Cell Nucleus and its parts.
    (b) Aleurone Grains.
    (c) Plastids.
3. NON-NITROGENOUS CONTENTS. Including Starch, Amylodextrin, Fatty oils and fats, Calcium Salts, Sulphur, etc.
4. CELL SAP. Containing:
    (a) Organic Substances in Solution. As Inulin, Hesperidin, Asparagin, Leucin, Tyrosin, other Glucosides, Alkaloids, Sugar, Mucilage, Tannin, Bitter stuffs, Coloring matters, Ethereal Oils, Resins, Gums, Rubbers, Plant Acids, etc.
    (b) Inorganic substances in solution. Salts of Sodium, Potassium, Lithium, etc.

I. THE UNFORMED NITROGEN-OUS CELL CONTENTS.

The main constituent is the Cytoplasm of the cell body. It is generally viscid in consistency, alkaline in reaction, and fills the cavities of very young cells. After the cells commence to grow portions of the Cytoplasm are consumed in the building up of the plant, and small vacuoles appear in it; these grow larger as the plant increases until the cytoplasm may make only a thin covering on the inside of the cell wall. In this condition the term Primordial Utricle has been applied to it, and the vacuoles may become filled with a transparent substance called the Cell Sap, which may come to occupy the major portion of the cell body. In still older tissues, as in woody stems, the cytoplasm may disappear entirely and the cell cavity may become filled with the cell sap, or be empty and dead, in which case it serves as a mechanical support only.

(Fig. 9.) Interspersed through it are many small granules, the Microsomes, whose function is still far from being understood.

The Cytoplasm is a very complex body, but it will not be in place here to more than outline its main characters.(a)

It is the seat of the active life processes of the plant and represents the results of the active metabolism of the plant tissues.

2. FORMED NITROGENEOUS CONTENTS.

(a) The Cell Nucleus.—This is a most important member of the cell body. It is generally immersed in the cytoplasm and lies, as a rule, to one side of the center. It is to be found in young growing tissues, but in older tissues it may be absent. The pharmacognocist rarely finds the nuclei in the tissues of drugs that he is called upon to investigate. Nuclei are generally single in plant cells, but they may be increased in number, as is to be found in the bast cells of hops and nettles and in some of the lacticiferous vessels of poppy and euphorbia, and in many of the lower orders of plant life, such as the Algæ and Fungi. Nuclei are generally small. The structure of the nucleus is very complicated. In general it consists of a nuclear membrane, quite delicate, which surrounds the plasma of the nucleus. In this plasma are a number of substances as yet little understood. The most important are those known as Chromatine and Pyrenine, of which latter substance the nucleolus, a minute granule found in the nucleus, is composed.(a)

In the nucleus the important process of Cell Division has its origin. This

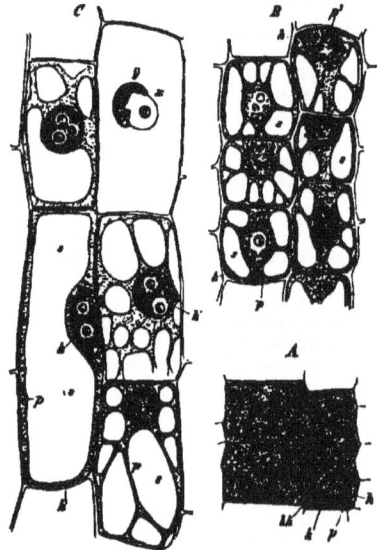

Fig. 9.—Parenchyma-cells from the central cortical layer of the root or *Fritillaria imperialis*; longitudinal sections (x 550). *A* very young cell lying close above the apex of the root, still without cell-sap. *B* cells of the same description about 2 mm. above the apex of the root; the cell-sap *s* forms separate drops in the protoplasm *p*, behind which lie walls of protoplasm; *C* cells of the same description about 7—8 mm. above the apex of the root; the two cells to the right below are seen in a front view; the large cell to the left below is in section; the cell to the right above is opened by the section; the nucleus shows under the influence of the penetrating water. a peculiar appearance of swelling (*x, y*).

process will be summarized under Growth.

(b) Aleurone Grains. These form the reserve nitrogenous materials of the plant in an analogous manner to the

---

(a) The student is referred to Reinke. Studien über das Protoplasma; Hertwig. Die Zelle und die Gewebe. 1892: for a more complete statement of what is known about this interesting substance.

(a) Hertwig. Die Zelle und die Gewebe.

starch grain, which is the great reserve for carbohydrate material of the plant. Aleurone grains vary considerably in size and shape, from round to elliptical, egg-shaped, or some times crystal-like in form. (a) In general they are colorless, sometimes brownish, greenish or yellowish. They are found abundantly in seeds and especially in oily nuts, often making up from 10 to 25 per cent. of such nuts as the almond and Brazil nut.

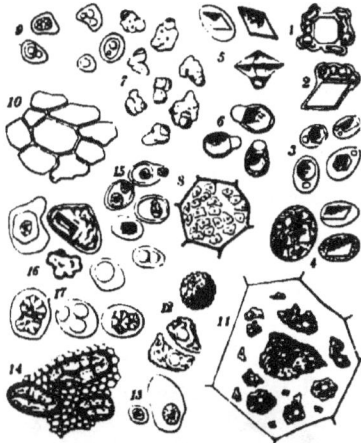

Fig. 10.—Aleurone grains from the seeds of 1 and 2—Bertholletia excelsea. 3. Ricinus communis, acted upon by water. 4. Elaeis guindensis. 5, Myristica fragrans. 6. Cannabis sativa. 7. Datura Stramonium. 8. Gossypium spec. 9 and 10, Cydonia vulgaris. 11, 12 and 13, Amygdalus communis. 14. Phaseolus vulgaris. 15. Coriandrum sativum. 16. Vitis Venifera. 17, Foeniculum (Tschirch) reduced.

(a) Morphologically the Aleurone grain consists of 1. the Membrane. 2. The Mass of the Grain. 3. The Inclusions. These may be crystalloids, globoids, or oxalate of calcium crystals.

(c) THE PLASTIDS. These bodies in great part are the active builders of plant tissues. They are universally distributed throughout the vegetable kingdom, failing to any great extent only in the fungi. We distinguish two types:

1. Leucoplastids. White or colorless.
2. Chromoplastids. Variously colored. Green, blue, yellow, red, etc.

1. The Leucoplasts are small, colorless albumenoid bodies widely distribut-

(a) For methods of staining the aleurone grains, see Zimmermann, Botanical Microtechnique.

ed in plants. They are the first steps in the building up of the Chlorophyll grain and through this to the starch grain. They are quite minute in size, and are most abundant in roots, especially in those roots that not only are storehouses for starch, but which also manufacture some, as Iris.

2. The Chromoplasts include a number of diverse elements, the most important one of which is the Chlorophyll grain. The Chlorophyll grains are the main organs of assimilation of the plant; by them the active energy of the sun's

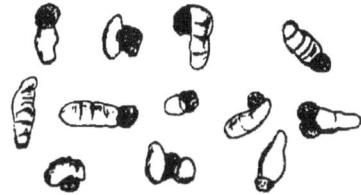

Fig. 11.—Leucoplastids in Rhizome of Iris, in process of forming starch. (Tschirch.)

rays is converted into potential energy and is stored up in the starch grain which the Chlorophyll grain is making at the same time from the carbon dioxide of the air and the water that comes from the root.

In many of the lower forms of plant life the chlorophyll grain is irregular in shape, but in the higher forms from the mosses upward they are rounded or slightly oval. In size they vary from 3 to 11 micro. m.

The structure consists of a protein body substance similar in all respects

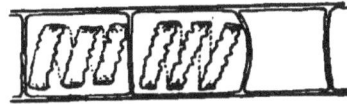

Fig. 12.—Cells of an alga showing spiral band-shaped chlorophyll grains.

to a leucoplast. This is somewhat spongy and is infiltrated with the coloring matter, which is generally green; the whole is often enclosed in a delicate albumenous membrane.

While the chlorophyll grain is usually found only in the vegetable kingdom, it is not exclusively its own, for there are a number of the lower animals, Protozoa, in which a certain amount of chlorophyll is found.

Many of the herbs and trees undergo wonderful color transformations which are, in the main, due to chemical modification of the chlorophyll; these modifications are of a very intricate nature which is but little understood. The brownish green is said to be due to the acid Phyllocyanin; it is the commonest coloration met with in the dried drugs of the market.(a)

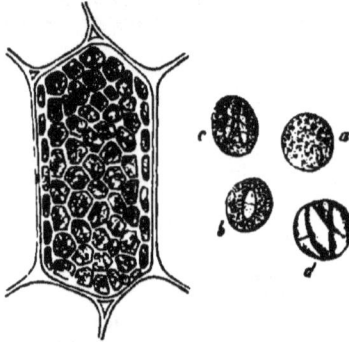

Fig. 13.—Chlorophyll grains E. A cell filled with chlorophyll grains. a Chlorophyll grain, b. c. d. starch inclusions in grains. (Tschirch).

The chlorophyll grain often contains Inclusions; these may be:

(a) Formed. Starch, the most important; Protein Crystalloids, rarely.

(b) Unformed. Fatty oils, coloring matters and a number of minute and little understood materials.

The coloring matters of the lower orders of plant life, as Diatoms, Marine

Fig. 14.—Chromatophores from flowers of Tropaeolum and from fruit of Capsicum. (Tschirch).

and Fresh Water Algæ that are other than green, are mainly mixtures of chlorophyll and other coloring materials. Chondrus crispus, our only official alga, is reddish to purple, and its coloring

(a) See Tschirch, p. 58.

matter, like many of its kind, is composed of chlorophyll and Phycoerythrin.

Other Chromoplasts. The numerous coloring matters of flowers and fruits are included under this comprehensive head. These coloring matters for the main may be considered as dissolved in the cell sap, though this is not the absolute rule by any means. In general it has been observed that the oranges and yellows are usually bound up with the plastids, and the blues and reds are in solution in the cell sap. These coloring matters when organized enough to have any form are quite irregular; triangular, circular, oval and polyangular forms are found in the same plant.

Bacterioiden. Within the past ten to fifteen years there have been discovered upon the roots of many plants, the family Leguminosæ in particular, peculiar tubercular swellings which have been found to contain small bacteria-like bodies. The function of these bodies, it has been inferred, is to take in nitrogen for the plant, and numerous investigations

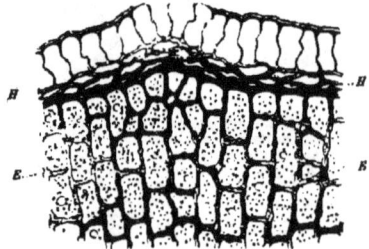

Fig. 15.—Aleurone grains in seed of Sabadilla H. Seed coats. E. Endosperm with oil drops and aleurone grains. (Vogl. reduced).

of the past few years seem to point in that direction. The subject is in much controversy, however, and time alone can give us the correct solution of the presence of these bodies. (a)

3. THE NON-NITROGENOUS CELL CONTENTS. The most important of these are the Fatty Oils, Fats, Starch, Amylo-dextrin, Calcium salts and Sulphates.

Oils and Fats. These are generally found either in the protoplasm or the cell body or in the cell cavity, in the

(a) Tschirch. Bericht. des Deut. Botan. Gesell., vol. V., p. 58. The literature up to that time is there given.

form of minute drops. Sometimes they are found in a more or less solid condition, irregular in shape. They should be considered as reserve products, sometimes replacing starch and sometimes in combination with starch, serving as nutriment for the young plant after germination. They are, therefore, most often found in the seeds and fruits of plants, less often in the stems or roots.

They are present in Senega, Gentian, Glycyrrhiza, Flos. Tiliæ and Chamomile, in most pollen grains and in the stigmas of Crocus, also in the spores of Lycopodium and in Ergot. Chemically they are ethers of glycerine formed with the fatty acid series, or acrylic series.

Starch is found throughout the vegetable kingdom in large quantities and two types are to be distinguished, Assimilated Starch and Reserve Starch. The first has been spoken of under the paragraphs upon the Leucoplasts and the Chromoplasts. The Reserve Starch is the starch of assimilation which has been dissolved and has passed through the leaves and into the bark, where it is called transitory starch and is in small grains 2-5 mm. in diameter. From here it passes on and is found stored up in the roots and stems, tubers and seeds to serve as a reserve food product. This reserve starch is generally in larger-sized grains measuring from 30-200 mm. in diameter.

There are numerous classifications of the starch grain; the most important of these are to be found in Wiesner's Mikroskopische Technologie and in Nägeli's Stärke Koerner. I here append the classification of Vogel:

A. Granules simple, bounded by rounded surfaces.
  I. Nucleus central, layers concentric.
    a. Mostly rounded or from the side lens shaped.
      1. Large granules, .0396-.0528 mm. Rye starch.
      2. Large granules, .0352-.0396 mm. Wheat starch.
      3. Large granules, .0264 mm. Barley starch.
    b. Egg shaped, oval, kidney shaped. Hilum often long and ragged.
      1. Large granules, .032-.097 mm. Leguminous starches.
  II. Nucleus eccentric, layers plainly eccentric or meniscus shaped.
    a. Granules not at all or only slightly flattened.
      1. Nucleus mostly at the smaller end, .06-.10 mm. Potato starch.
      2. Nucleus mostly at the broader end, or toward the middle in simple granules, .022-.060 mm. Maranta starch.

    b. Granules more or less strongly flattened.
      1. Many drawn out to a short point at one end.
        a. At most .060 mm. long. Curcuma starch.
        b. As much as .132 mm. long. Canna starch.
      2. Many lengthened to bean shaped, disk shaped or flattened; nucleus near the broader end, .044-.075 mm. Banana starch.
      3. Many strongly kidney shaped; nucleus near the edge, .048-.056. Sisrynchium starch.
      4. Egg shaped; at one end reduced to a wedge, at the other enlarged; nucleus at the smaller end, .05-.07 mm. Yam starch.
B. Granules simple or compound, single granules or parts of granules, either bounded entirely by plain surfaces, many angled, or by partly rounded surfaces.
  I. Granules entirely angular.
    1. With a prominent nucleus. At most .0096 mm. Rice starch.
    2. Without a nucleus. The largest .0088 mm. Millet starch.
  II. Among the many angled, also rounded forms.
    a. Few partly rounded forms present, angular form predominating.
      1. Without nucleus or depression, very small, .0044 mm. Oat starch.
      2. With nucleus or depression, .0132-.0220 mm.
        a. Nucleus or its depression considerably rounded; here and there the granules united into differently formed groups. Buckwheat starch.
        b. Nucleus mostly radiatory or star shaped; all the granules free. Corn starch.
    b. More or less numerous kettledrum and sugarloaf-like forms.
      1. Very numerous eccentric layers; the largest granules .022-.0352 mm. Batata starch.
      2. Without layers or rings, .08-.022 mm.
        a. In the kettledrum-shaped granules the nuclear depression mostly widened on the flattened side, .008-.022 mm. Cassava starch.
        b. Depression wanting or not enlarged.
          aa. Nucleus small, eccentric, .008-.016 mm. Pachyrhizus starch.
          bb. Nucleus small, central or wanting.
            aaa. Many irregular forms, .008-.0176 mm. Sechium starch.
            bbb. But few angular forms; some with radiatory nuclear fissure, .008-.0176. Castanospermum starch.
C. Granules simple and compound, predominant forms egg shaped and oval, with eccentric nucleus and numerous layers, the compound granule made up of a large granule and one or more relatively small kettledrum-shaped ones, .025-.068 mm. Sago starch.

Amylo-dextrin is not found free in nature, but in combination with starch, re-

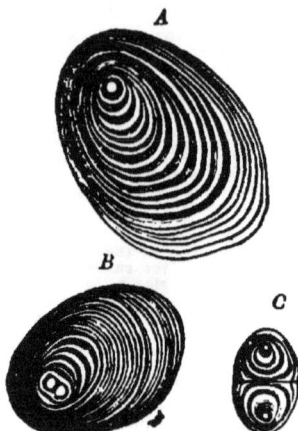

Fig 16.—Starch grains of Potato. (Sachs.)

more or less needle-shaped; hexagonal forms are also found, as well as mono-symmetric and octagonal forms. In some plants, Cortex Cinnamomum, Bella-donna, Dulcamara, the crystals exist as crystalline "meal" that can only be distinguished by means of the polari-scope. Special forms have received the

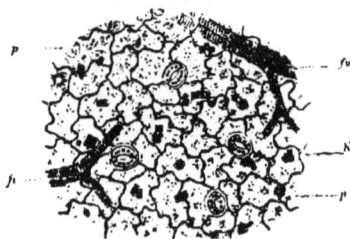

Fig. 17.—Upper surface of leaf of Hyoscyamus, showing oblong crystals. *K*. Also vessel bundles *f. v.* (Vogl.)

acting in this combination with iodine a violet red or reddish brown.

Mineral Salts. Salts of calcium, oxalate, carbonate, sulphate and phosphate.

Calcium oxalate crystals are widely distributed in plant cells, being found in all parts of the plant save the cells that functionally are used for purposes of conveying plasma materials. The commonest form is that of crystals occurring in clusters or singly. They are

name of Rhapides, or needle-shaped; Quadrangular, or rhomboidenal, and Agglomeratur, or clustered crystals.

Calcium carbonate is less often found than the oxalate. It may be in incrustations covering cell walls or in the form of cystoliths, which are mulberry-like masses; these are to be found in Cannabis, Humulus and Ficus in abundance

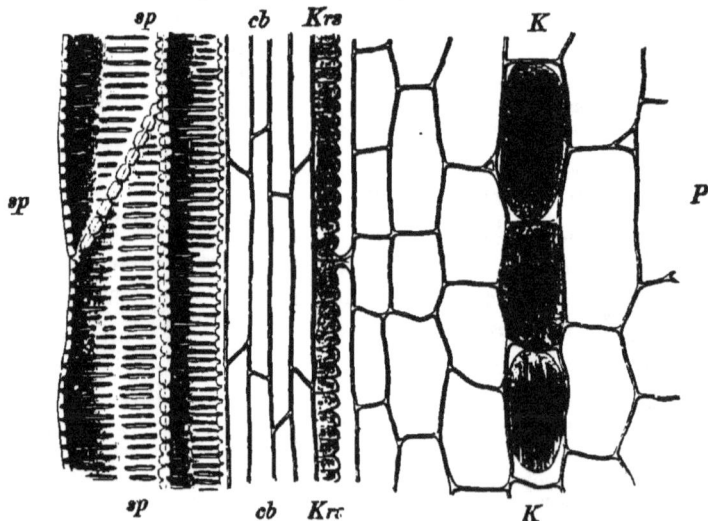

Fig. 18.—Rhizome Veratri. Longitudinal section of portion of root. *P.* parenchyma, *K.* acicular crystal sacs, *Krs.* Endodermis, *Sp.* Ducts. (Vogl.)

on the epidermal tissues, especially those of the leaves. The inner portion of the cystolith is free from $CaCO_3$.

Microchemically, acids have no effect upon calcium oxalate, while they dissolve the crystals of calcium carbonate with a formation of gas, bubbles of which may be seen by means of the microscope.

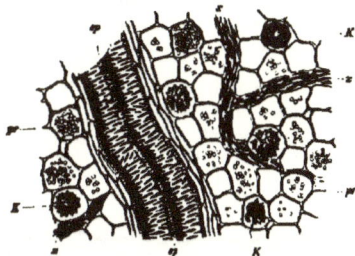

Fig. 19.—Rhizome Rhei. Longitudinal section, *sp.* Vessels, *K.* Crystal sacs, *pr* starch conducting parenchyma. *X.* Obliterated parenchyma.

Calcium sulphate and phosphate are rarely found in crystalline condition. The phosphates are a constant ingredient of plant ash, but are seldom seen in a crystalline form save in a few plants. Euphorbiaceæ.

THE CELL SAP. It has already been pointed out that in the vacuoles and other spaces left by the retreating protoplasm in the cell cavity, and pene-

Fig. 20.—Cross section leaf of Hemp, showing *C.* Carbonate of Calcium crystals (crystoliths), *or.* oil glands, *sp.* Stoma. *Sch.* mesophyll. (Tschirch).

trating at times the cell wall and the protoplasm, there is a watery material with an acid reaction. This goes by the name of the cell sap. It contains in solution many materials both organic and

inorganic. The soluble organic substances include Inulin, Hesperidin, Asparagin, Aloin, Sugar, Alkaloids, Glucosides, Acetates, inorganic salts, etc.

Inulin is an isomer of starch, and, like it, is a reserve material. It is found extensively in the roots of plants of the Compositæ, in which it varies from 1–50 per cent. Inula Helenium has 44 per cent., Taraxacum 20 per cent., Pyrethrum 50 per cent. Microchemically it is to be distinguished by its insolubility in alcohol and glycerin. If the fresh roots be placed in these solutions the Inulin will precipitate out in the form of sphæro-crystals, which fill the entire cell. Molisch's test is also of value. A-

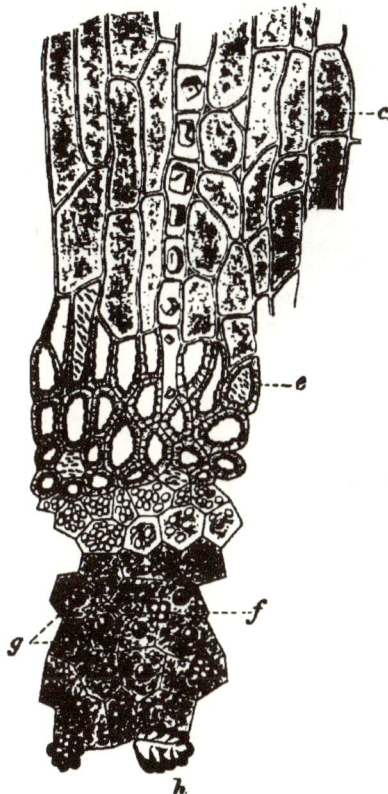

Fig. 21.—Cross section of Aleppo Gall, showing *g.* Tannin particles. (Hartwich).

naphthol, in 15–20 per cent. alcoholic solution, is added to the section on the slide, and after 2 to 3 drops of sulphuric acid have been added, if there be present cane sugar, milk sugar, glucose, levulose, maltose or inulin, the section will become violet in three or four minutes. Thymol used in the same way

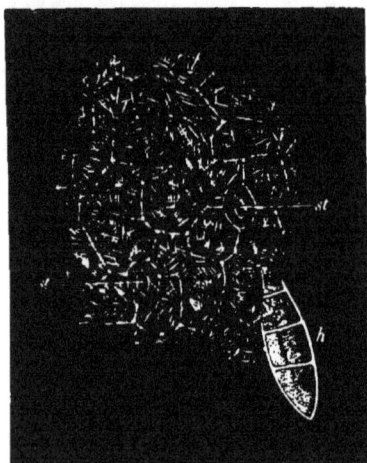

Fig. 22. – Underside of Epidermis of Duboisia, treated with warm potassa showing precipitate of alkaloid, Duboisin (Hyoscyamin). × 250. (J. Moeller).

gives a carmine-red reaction. Molisch's test is, however, not exclusive enough. Orcin in a saturated alcoholic solution boiled with hydrochloric acid gives an orange red solution if inulin be present.

Asparagin, Leucin, Tyrosin can only be mentioned with aloin, quassin, santonin and other complex chemical compounds, which are often included under the loose title of neutral principles. Asparagin is of interest on account of its relation to the formation of albuminoid materials, it being a sort of transition product in the formation of the nitrogenous materials.

Sugar in its many forms is commonly found in the cell sap. As thus found its most important forms are Dextrose, Levulose and Saccharose. Less often Mycose, Melitose, Synanthose, Inosite and Milk Sugar have been described as present.

Microchemically, Trommer's test and Fehling's are to be recommended. It should be borne in mind, however, that the tests for sugar will react to many other substances found in plant tissues, such as glucosides, etc.

Glucosides stand in close relation to sugar. They are built up of complex molecules of carbon, hydrogen, oxygen and sometimes nitrogen, which possess the common property of being converted into glucose and other allied products by the addition of weak acids or inorganic ferments. Some of the more common glucosides are Salicin, Solanin, Hesperidin, etc.

Tannins are complex compounds. They are found extensively in the barks of plants and throughout the leaves. They occur either in solution in the cell, or in the form of small grains, or even in incrustations of the seed walls.

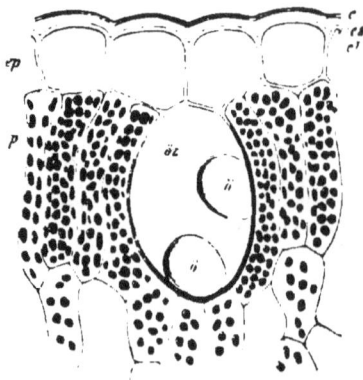

Fig. 23.—Cross section of Sassafras leaf showing o. oil drops in oz. suberized oil cell. ep. epidermis. p. Palisade cells. (Tschirch).

Alkaloids. Under this heading is generally included a large number of natural basic compounds which are nitrogenous and which give certain characteristic chemical reactions. They are probably products of destructive metabolism and are found in the cell sap. They are generally rich in the meristematic tissues, but their exact locality is still in doubt. It is probable that in an unaltered condition alkaloids are not to be detected by the microscope. By means of microchemical tests, however, many have been isolated and studied. In many cases the active medicinal principles depend upon alkaloids, as in Morphine, Strychnine, Physostigmine, etc. Many of the alkaloids are highly poisonous, but there are a number that are inert.

Resins, Oleo-Resins, Gum-Resins, Ethereal Oils and Balsams, all exist singly or in combination, either in the cell sap or in special passages, excretory passages or glands distributed throughout the plant body. Medicinally they form an important class of products.

The Gum-Resins are often found in the so-called milky juices of the plants. Balsams generally contain either Benzoic or Cinnamic acid. They are very complex mixtures. Caoutchouc is found quite extensively in milky saps.

The Volatile Oils are widely distributed and are to be considered as secretions. They may be in the cell sap or may exist in special glands or intercellular spaces.

Ferments form an interesting class of nitrogenous plant products. In the germinating seed they are of vital importance, converting many products for the use of the young plant. Diastase, Papain, Emulsin and Myrosin are among the most important.

Plant Acids. The aromatic plant acids are widely distributed. Some are important to the constructive metabolism of the plant body, but their offices are as yet not fully known, while others are employed in destructive metabolism, or katabolism.

CHAPTER IV.
THE CELL WALL.

The cell wall is the limiting membrane of the cell. When present, which is the rule, it generally consists of a mixed substance called Cellulose, which has been produced by the activity of the protoplasm of the cell. At first it is quite delicate, but as growth continues certain modifications take place which may be studied *as external* under the morphology of the cell wall, for they deal with the markings, thickenings, etc., and have nothing to do with the internal changes or chemical modifications.

Morphology.

In the young developing cells the cell wall appears like a thin veil, which grows both in thickness and in surface. How the growth in thickness takes place is not yet perfectly understood. (a)

Whether the cytoplasm which is found in the cell wall is transformed into cellulose, or whether its surface layers are added externally, is a subject of controversy. The truth probably is that both processes are at work at the same time.

This growth in the cell wall may take place centrifugally, from the inside outward, as in the walls of spores and pollen grains and epidermal cells; or centripetally, from the outside inward, as in stone cells, and in spiral and annular ducts.

The growth in thickness gives rise to many irregularities which will be studied later under the subject of pores and

(a) Zimmermann, Pflanzenzelle, p. 154. Strassburger (Zell und Zelle-bildung).

markings. The growth on the surface also gives rise to other irregularities which are determined by the amount of pressure exerted upon the growing cell. Cells, when uninfluenced by outside forces, tend to grow in a spherical shape, but this condition is rarely found in the higher forms of vegetable life, and unequal pressures produce cells of almost all degrees of irregularity, hexagonal, cubical, pyramidal, star-shaped, brick-shaped, flattened, etc.

The degree of variety is enormous, and in many cases cell shapes are so peculiar that they afford an excellent means for the identification of certain drugs.

Chemistry.

The chemistry of the cell wall is an intricate problem. The student is referred to Zimmermann, l. c., for a clear exposition of what can here be hardly more than touched upon.

Cellulose, which is generally taken as the substance of the typical cell wall, is a carbohydrate, with the empirical formula, $C_6H_{10}O_5$. Its micro-chemical reactions will be considered in a special chapter.

Botanists have considered five modifications of the typical cell wall, which are the results of the incrustation or saturation of certain chemical substances upon or in the walls of cellulose. These are:

1. The mucilaginous modification.
2. Lignification.
3. Suberization.
4. Cutinization.
5. Mineralization.

(1) The mucilaginous modification is seen in Chondrus crispus, and in many seeds, as in Quince, Flax, Mustard, Cassia fistula, etc. Such walls have the property of swelling to a great extent upon the addition of water, and all degrees may be traced, from pure cellulose, which is insoluble in water, to pure gum or mucilage, which may be entirely soluble.

(2) Lignification is supposed to be due to an incrustation of Lignin or woody substance upon the cellulose of the cell wall. This lignin is a hypothetical substance which gives a number of certain micro-chemical reactions, but its nature is far from being understood. It gives to cell walls a toughness and elasticity that are characteristic of wood and bast fibres.

(3) Suberization is seen in cells of cork; it renders them elastic and impermeable to water. It has been taught that it is due to a deposit of fat-like substance upon the cellulose of the wall; but many investigators have claimed that there is no cellulose in cork cell walls, and that suberin is a separate chemical substance.

(4) Cutinization is the process that takes place in the epidermal tissues of the plant, generally in the leaves and the outside covering of fruits. The properties of Cutin are similar to those of Suberin. Micro-chemically they agree.

(5) Mineralization. Mineral salts, particularly carbonates and silicates, are sometimes deposited in the walls of certain plant cells, as in grasses, horse tails, etc. Their function seems to be purely mechanical.

CHAPTER V.
THE GROWTH OF CELLS TO FORM TISSUES.

If the evolutionary point of view be kept in mind, it becomes necessary to trace the growth of highly-complicated tissues, as found in the higher plants, by means of the simpler forms of the lower plants (Phylogeny), or through the individual stages of growth from the fertized ovule to the fully developed plant (Ontogeny). Owing to the gaps in Paleobotany the Phylogenetic relationships of plants are difficult to trace, and the study of plant embryology affords a simpler clue to the origin of complex structures.

CELL DIVISION. The two main types of cell division have been described, the direct and the indirect.

1. The Direct method consists of the splitting of a cell into two, without any nuclear mechanism. It is confined mainly to the lower forms of vegetable life.

2. The Indirect method (Mitosis, or Karyokinesis), affords an interesting and complicated study, which is mainly centered in the nucleus. It has been pointed out that the nucleus consists of several protoplasmic substances, the most important of which was the Chromatin. This is generally scattered about the nucleus in the form of minute granules, but when the process of karyokinesis is about to take place it is noticed that the chromatin commences to arrange itself in irregular lines or a "skein." At the

same time the Centrospheres, or "Attraction Spheres," two very minute bodies at one side of the nucleus, outside of the nuclear membrane, separate and wander to opposite sides of the nucleus. These spheres seem to exert some influence upon the chromatin threads (Chromosomes), for they are soon found arranged in the shape of a spindle reaching toward the centrospheres, and are apparently arranged along radiating lines that stretch from one attraction sphere to another. (Achromatic Spindle.)

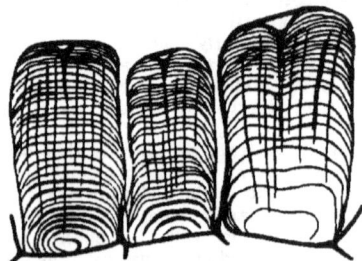

Fig. 74. — Epidermis cells of seed of Sinapis alba, showing mucilaginous modification after treat with water. (Tschirch).

The second stage begins with the disappearance of the nuclear membrane and the longitudinal division of the chromosomes. The chromosomes then come closer together in the center and divide transversely. (Metakinesis.) After

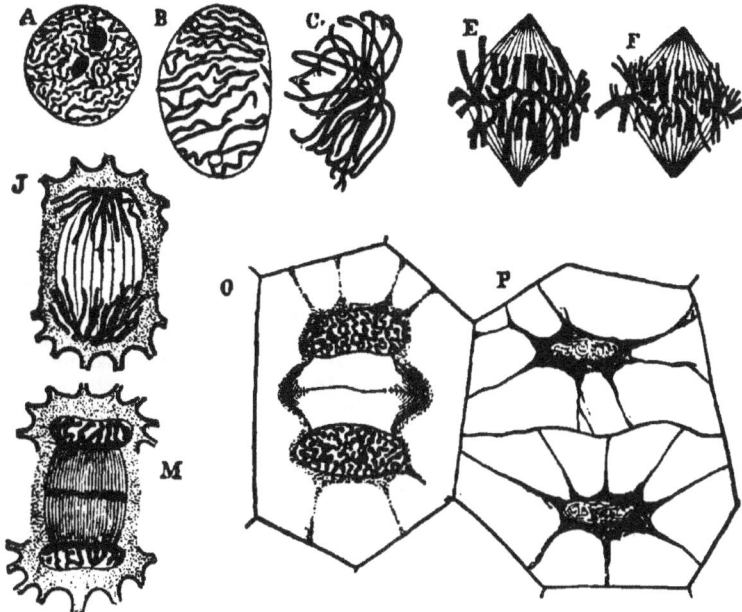

Fig. 25.—Process of Karyokinesis. *A* resting stage. chromatin in dots, lines just forming. *B.* stage of "skein" chromatin in lines. centro spheres at opposite poles *C.* solution of nuclear membrane. *E.* formation of achromatic spindles. the threads becoming arranged along them, *F.* longitudinal and transverse division of chromosomes completed. *J.* formation of "daughter nuclei". *M.* daughter nuclei formed. beginning of separation of the protoplasm. *O. P.* completion of process by division of protoplasm and formation of middle of lamella.

transverse division the two halves move to the poles or attraction spheres, and soon a transverse line is seen stretching across the protoplasm of the cell. This soon develops into a new cell wall. The two new cell walls thus formed, each containing its own nucleus, then round out and resemble the parent cell.

This process, hastily sketched, is constantly taking place in all growing animal and vegetable tissues. The details of the process are inexhaustible.(a)

It is by a continual cell division of this type that the tissues about to be studied have been formed. Originally simple and homogeneous in the embryo, they become complex and heterogeneous in the mature plant.

Classification.

A satisfactory classification of plant tissues is difficult. as anatomical structure and physiological functions vary so widely. The following classifications

(a) Strassburger Zell und Zell theilung. Guignard, l. c.

look at plant tissues from four separate standpoints.

1. Plant tissues may be composed of either

(a) Growing, or meristematic, tissue.

(b) Permanent tissue.

2. Plant tissues may be classified according to the shapes of the cells. This is unsatisfactory, but as the names of the variously-shaped cells are extensively used, they must be given.

(a) Equal diameter cells, generally thin-walled. Parenchyma.

(b) Unequal diameter, generally elongated cells, with thickened walls. Prosenchyma.

3. A third classification groups plant tissues into systems that are based upon their anatomical position.

(a) The external epidermal tissues.

(b) The internal fibro-vascular tissues.

(c) The ground, or filling, tissues.

4. A more comprehensive classification looks at plants from a point of view both anatomical and physiological.

This is the classification that will be here followed.

(A) Formative tissues. Meristem.

(B) Protective tissues.

   1. Epidermal tissues. Epidermis, cork.

   2. Skeleton, or Mechanical, tissues. Bast fibres, Libriform, Sclerotic, or Stone cells.

(C) Tissues of Nutrition.

   1. Absorption. Root hairs, Haustoria.

   2. Assimilation. Palisade cells and other leaf tissues.

   3. Respiration and Transpiration. Stomata, Lenticels.

   4. Conduction. Ducts and Sieve tubes.

   5. Secretion and Excretion.

(D) Reproductive tissues.

CHAPTER VI.
FORMATIVE TISSUES.

Meristematic tissues are tissues that have the power of forming new cells, hence the power of growth is said to be confined to them. In the higher plants the points of meristematic growth are to be found only at the apex of stems both apical and lateral, the ends of roots, the cambium layer, the phellogen layer and the tips of leaves and flower structures. Under special circumstances other parts may take on meristematic parts, as, for instance, when a plant is injured new tissues, mainly of a corky character, may be formed in large quantities.

Meristematic cells are generally thin-walled, square, angled, and filled with protoplasm and provided with nuclei. In some very rapidly growing meristems the cell walls may be absent in the early stages of growth.

Meristematic tissue may be primary or secondary. Primary meristem is found

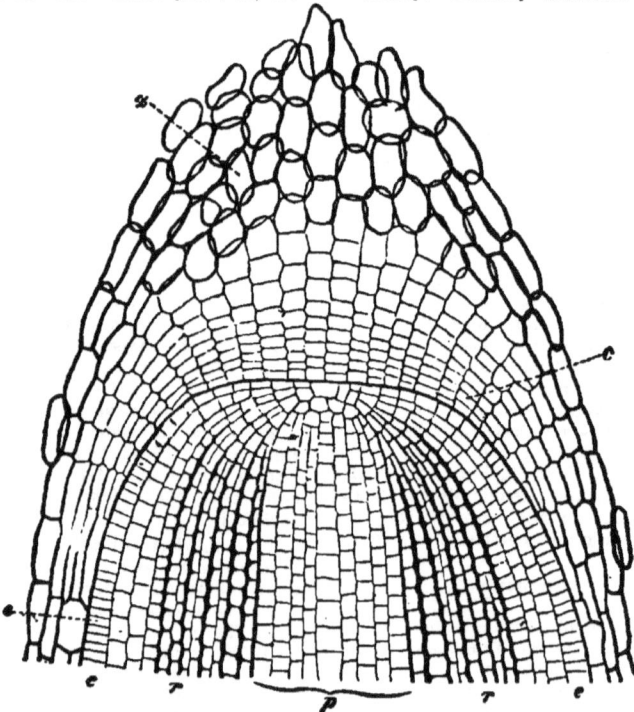

Fig. 28.—Apical growths. e. dermatogen. r. plerome. p. periblem.

only in the young embryo and in the tips of growing plants, and all the rest is secondary in its formation.

It has been extensively taught that in the apical region the primary meristem gives rise to three layers of cells from which the more complicated tissues spring. The outer layer is called the Dermatogen and gives rise in general to the epidermal system of tissues. Beneath this lies the Plerome, which gives rise to the fibro-vascular system, and within is the Periblem from which the ground tissues arise. This grouping is adopted merely for the sake of convenience, as it is not universal in plants themselves.

Much of the tissues that are built up by the secondary meristem becomes permanent in form and makes the "Permanent tissue" of a former classification. Drugs consist in general of permanent tissues, but in the following meristematic tissues may be found: Cambium in Dulcamara stems, Radix Angelica and Althæa and Secondary Meristem in Jalap tubers. Apical cells may be found in many herbs after treatment with dilute ammonia or alcohol. Phellogen is to be found in oak bark, Inter fascicular Cambium in Ricinus stem, and Primary Cambium on the radicles and cotyledons of nearly all the seeds, such as Capsicum, Croton, Ricinus, Sinapis, etc.

CHAPTER VII.
PROTECTIVE TISSUES.

The protective tissues of the plant consist of the Epidermal and Mechanical Systems.

EPIDERMAL SYSTEM. The epidermal system is the outermost layer of the plant and completely covers every portion of it, protecting it from outside mechanical injury and from too rapid evaporation and loss of heat.

In young plants the epidermal system consists of the epidermis with its two modifications, the hairs, or Trichomes, and the Stomata. In many stems a year or so old secondary changes take place which lead to a new formation of tissues which form the secondary epidermis. The epidermis, or skin, is the superficial layer in roots, stems and leaves, and consists of very variable cells. In the

Fig. 27.—Belladonna leaves. Under side showing epidermal cells, here slightly striped. *st*, A, and *fv*, stomata, hair, and vessels respectively. (Vogl).

official cryptogams Cetraria and Chondrus, the epidermis is not a true epidermis, but in Aspidium there is a well-

marked superficial layer of epidermal cells.

The epidermal cells are generally flattened and united to one another without intercellular spaces. They are usually elongated in the direction of the axis in the monocotyledons and are irregular in Dicotyledons. Often in seeds they are thicker than wide, but this is not the rule.

The outer wall of most epidermal cells is thickened by a deposit of cutin which renders it impervious to moisture and also serves to protect it against insect and fungus pests. This layer of cutin varies greatly in thickness and bears a direct ratio to the amount of transpiration that serves the purpose of the individual plant. In plants with leaves more or less horizontal the side exposed to the sun's rays is generally thicker. In tropical countries where the heat is excessive and the loss of moisture would be great the epidermis is much thicker than in temperate climates.

The modifications of the epidermis are:
In roots—Trichomes.
In stems—Trichomes, glands, stomata, water pores.

Trichomes are modifications of the epidermis and consist of outgrowths of a single cell. They may be very simple, hardly more than a slight projection, or very complex, forming many-celled, jointed and branched hairs. Their contents may consist of protoplasm with a cell nucleus, and sometimes crystals of lime salts.

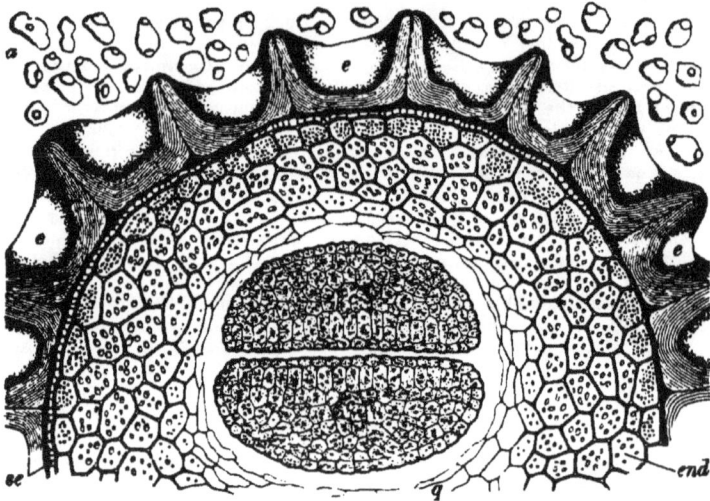

Fig. 28.—Hyoscyamus niger. Cross section of seed showing *e* enlarged and modified epidermal cells, *end*. Endodermis. *c*. Cotyledons. *a*. isolated aleurone grains. (Tschirch).

In roots they are generally simple and are called root hairs, are somewhat elongated and have very thin walls and are situated a little behind the root cap and

trichomes are of almost every conceivable variety. Some hold no secretion, and are called hairs in the narrow sense of the word, and others contain secre-

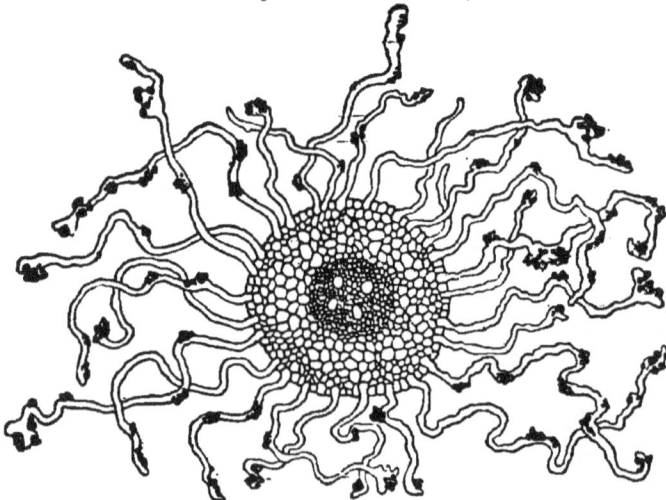

Fig. 29.- Root Hairs.

limited to a small area. Their function is mainly that of absorption of moisture holding mineral salts in solution.

In stems and leaves and flowers the

tions and are called glands. These hairs, or papillæ, as they are called when in floral petals, serve as a means of identification in Viola flowers, where they

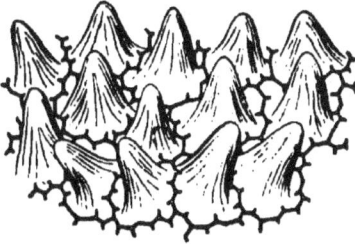

Fig. 30.—Simple papillae on petals of Viola tricolor. (Wiesner).

occur as short unbranched or branched, conical hairs. In the leaf of Cheiranthes the trichomes are flattened, one-celled or many-celled, scale-like hairs. For pharmacognostical purposes plant hairs are of great importance, as they

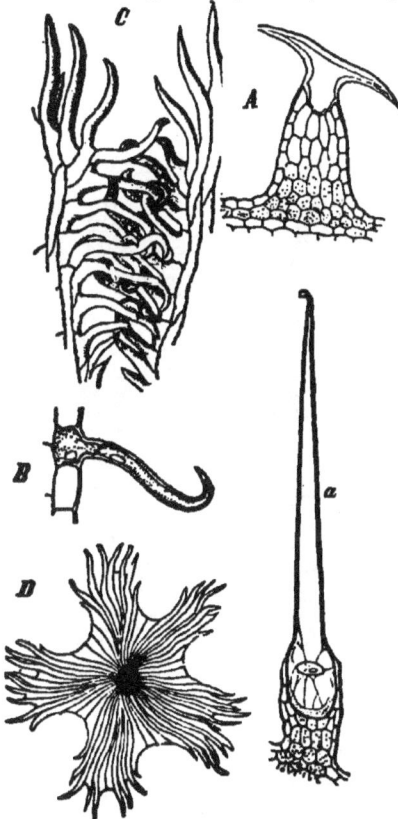

Fig. 31.—Variously shaped hairs.

form the diagnostic features by which certain ground drugs, especially those made from leaves, may be recognized.(a)

Glands are epidermal growths, generally various in shape, few to many-celled, and containing excretions. They will be discussed under the head of Secretory Tissues.

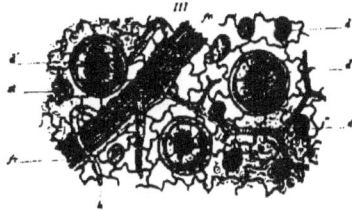

Fig. 32.—Leaf of Mentha piperita showing upper surface. *h*, simple hairs. *d*, glandular hairs. (Vogl).

Water Pores and Stomata are modifications of the epidermis, but as they make such a large part of the leaf structure and have special physiological significance they will be discussed under the subject of Transpiration and Assimilation.

Secondary growths of the Epidermis. In the stems of most plants the epidermis serves as a means of protection until about the end of the first year's

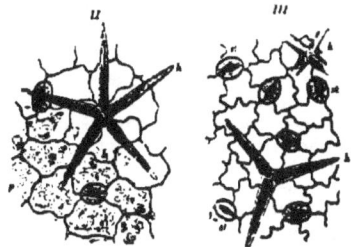

Fig. 33.—Upper II and under III. sides of leaf of Althæa showing star shaped hairs. (Vogl).

growth; after this the increasing diameter of the stem causes it to crack off in varying degrees, thus leaving the tissues beneath unprotected. Hence a secondary growth, called the Periderm, takes the place of the primary epidermis. This periderm varies greatly, but in the main it consists of two parts, meristematic

(a) Weiss. Die Pflanzen haare. Karstens Botanische Untersuchungen vol. I. DeBary. Comparative anatomy.

give the cork of commerce.

In many plants the phellogen layer makes an inner periderm as well as the outer one of cork; this is the case in Althæa, Glycyrrhiza, Ipecacuanha, Levisticum, Rhatany, Angelica, Senega, etc. In the inner periderm, whether of phello-

Fig. 34.—Glandular hair from leaf of Hyosoyamus niger. (Tschirch).

cells that aid in the formation of new cells, and the product of the meristematic tissue, namely, the cork.

The meristematic layer is termed the Phellogen Layer, and has its origin in the parenchymatic cells that lie just beneath the epidermis. These cells divide by tangential cross walls, the inner products of the division retaining the power to divide, while the outer products become cork cells.

Cork Cells are generally thin-walled and quite regular in shape, having the form of flattened tablets. They are brownish-red in color and give the characteristic reactions of suberin. The contents are mainly air, but often various coloring matters are found giving the cork its peculiar color. Although typically thin-walled, the cork cells may have thickened walls which may become hard and stony as seen in the plane tree. Cork cells vary in their number of layers from one cell deep, in the potato, to many feet in thickness in the oaks that

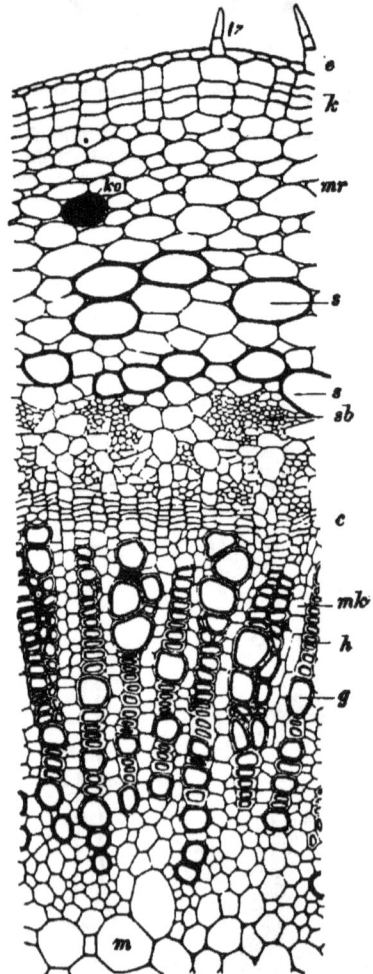

Fig. 35.—Cross section of young shoot of Cinchona Calisaya, showing the beginning of the Periderm. *k* phellogen layer forming cork. *e* epidermis with hairs still present, *mr* primary rind. *s* mucilage reservoirs: *sb* sieve tubes in the secondary cortex. *c* cambium, *mk* medullary rays, *m* pith. (Tschirch).

Fig. 36.—Cross section of false Cinnamon showing corky formations. (Moeller).

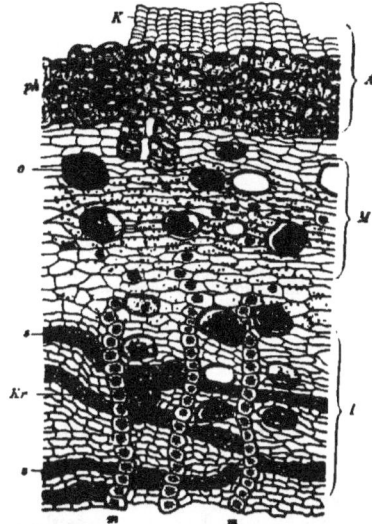

Fig. 37.—Cross section of bark of Canella alba.
A. outer bark, with *k* cork. *ph* phelloderm, *M*.
middle bark with suberized oil reservoirs and
crystals, *I*. inner bark with medullary rays which
contain crystals of calcium oxalate, *s*. obliterated
sieve tube elements. (Tschirch).

against internal and external pressures
both above and below ground.

The most important mechanical tissues
are Sclerotic Cells, Collenchyma, Wood
Fibres and Bast Fibres. The word
Stereom expresses these cells taken collectively.

genic or pericambium origin, there may
also be found collenchyma, sclerotic
cells, bast fibres, crystals, resin passages
and oil cells. When the entire periderm
is made up of such diverse elements it
has been termed bark.(a)

Pharmacognocists recognize three parts
in barks. The Outer, Middle and Inner
Barks. The Outer bark includes the
corky epidermal layers on the outside of
the stem; the Middle bark is what remains of a portion of the primary epidermis, it becomes quite complex and
contains a variety of substances, crystals, chlorophyll, tannins, oil, etc. The
Inner bark lies between the middle bark
and the cambium and is traversed by the
medullary rays of the phloem.

MECHANICAL TISSUES, OR THE
PLANT SKELETON. These tissues
are developed by the plant for the purpose of internal protection of the more
delicate growing and conducting tissues,

Sclerotic Cells. Sclereids. Stone Cells.
These are thick cells varying from iso-
diametric to oblong, round, oval or irregular in shape. They are strongly lig-

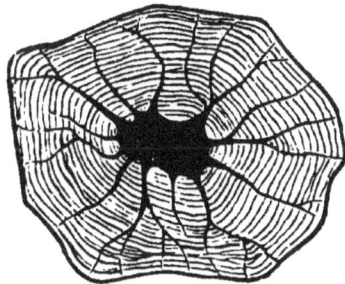

Fig. 38.—Stone cell from Pimenta, showing
stratification and branched pore canals.
(Tschirch).

(a) J. Moeller, Die Baum rinden, Berlin,
1882.

nified and have pitted walls. They are often yellowish in color, but not always. They may be distinguished from bast fibres by being shorter and having round instead of oval pores. In the fruit of Pears, Pepper, Pimenta, and barks of

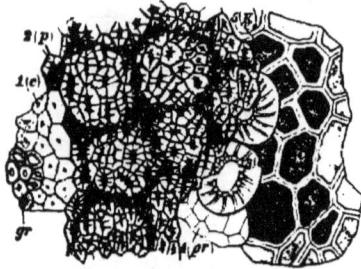

Fig. 39.—Surface section of Foenugreek seed, showing thickened walled stone cells. (Tschirch.)

Cinnamon and Oak the more regular types are to be found. In the seed coats of Physostigma, Abrus, Ricinus and Croton more elongated types may be found, while in tea leaves irregular types occur. Sclerotic cells are best isolated by means of Schultz's maceration fluid when they can be studied to better advantage.

Collenchymatic Tissues. These tissues consist of cells that are of an intermediary type. They often serve two functions, those of mechanical support and the transmission of fluids by osmosis. Hence their walls are thickened at the

angles, leaving the remainder thin and capable of transmitting fluids. By the juxtaposition of several angles there is formed a strong band which serves as a mechanical support. This is the mechanical tissue of young, growing parts, and is generally the first kind that is developed. It is absent in most Monocotyledons, but is widely distributed in Dicotyledons, being found in the stems and leaves and petioles of most young herbs. In official plants it is found in stems of Conium, Levisticum, Malva, Mentha, in leaves and petioles of Laurel, Mentha, Senna, Conium, Tussilago and in the flower stems of Malva, Sambucus, etc.

Bast Fibers. (Schlerenchyma in a

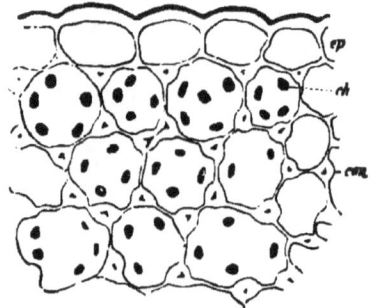

Fig. 40.—Collenchyma cells in stem of Umbelliferous plant. *p*. epidermis. *ch*. chlorophyll grains. *con*. collenchymatic thickening.

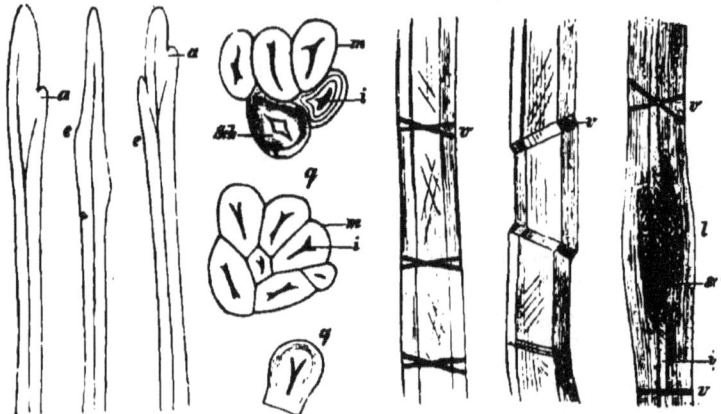

Fig. 41.—Bast fibres from Hemp. *q*. cross sections, *e*. ends, *v*. dislocations, *i*. lumen, *s*. fine striations. (Von Hoehnel).

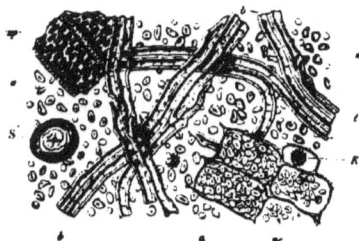

Fig. 42.—Powdered althaea root, showing, *b.* bast fibres, *sp.* Pitted vessel, starch etc. (Vogl).

narrow sense.) These are in general the longest and strongest cells of the plant They are usually much elongated

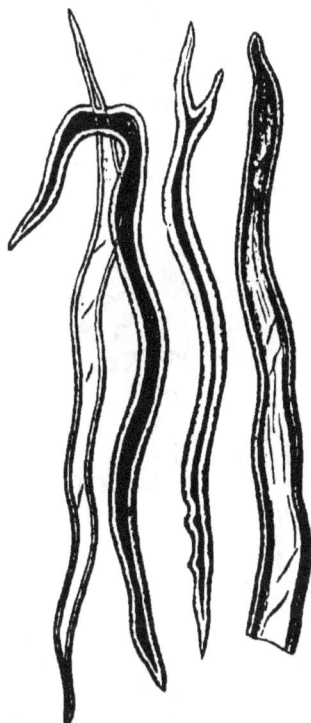

Fig 43.—Wood fibres or Libriform. (Tschirch).

and thick-walled, and have sharpened points and small, spirally arranged, slit-like pores. As a rule, the bast cells have no contents; the wall is generally lignified and the lumen filled with air.

The length varies greatly from ½ mm. to an inch or more, the proportion of length to breadth being about 10 to 1; often however, bast fibers are from 2,000 to 4,000 times as long as wide. (Fig. 41.)

The thickening of the cell wall varies greatly, being generally very unequal. In some plants the bast fibers are but slightly thickened, and have a wide lumen resembling wood fibers; at other times the lumen is entirely gone.

On cross section the bast fibers when alone are *more* or less rounded; but when collected into bundles, mutual pressure tends to make them angular, polyhedronal, or even apparently iso-diametric.

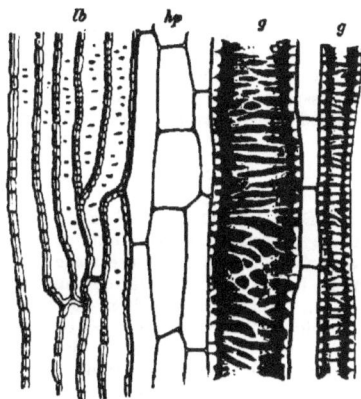

Fig. 44.—Longitudinal section of root of Inula Helenium, showing *lb.* libriform fibres with pores, *g.* vessels. (Tschirch).

The ends vary greatly; typically they are sharp and somewhat rounded; occasionally they are squarish, and sometimes branched, in which case, especially when short, they are hard to differentiate from some stone elements.

The slit-shaped pores are characteristic, especially in their left-handed spiral arrangement. (DeBary, l. c.)

The strength and toughness of the bast fibers are great. Experiments have shown that some of the stronger fibers will stand proportionately the same bending as steel without breaking, and that they can bear from one-quarter to one-half as great a pull as this metal, while they are from five to ten times as elastic, that is, they will allow of so much more proportional elongation.

Bast fibers are found mainly in the inner bark of most Dicotyledons, and are distributed throughout the stems of Monocotyledons, where they generally surround the fibro-vascular bundle as a protection. They, with wood fibers, are specially well adapted to support those plants which have to stand great strains and are few or absent in plants not similarly subjected, as in water plants and those leaves and stems that float or lie flat on the ground. (Fig. 42.)

A variety of bast, called soft bast, will be discussed under the conducting tissues.

Wood Fibers, Libriform, Woody Sclerenchyma, are different names given to the so-called woody parts of the plant, the greater proportion of which is made up of such fibers; they are in most respects similar to bast fibers, as the word libriform indicates. They are more slender and are neither as strong nor as elastic as bast fibers. In many forms the pores may be circular instead of slit-like. In stems of Dicotyledons they occur as elements of the Xylem inside the Cambium ring, while the bast fibers are found in the Phloem without. In many plants, however, it is impossible to distinguish them if isolated. (Fig. 43.) In some intermediate forms between wood fibers and tracheids border pores are to be found.

Woody fibers contain in general nothing but water or air, and in some cases some shrunken plotoplasm remains.

Fibrous Cells, or Intermediate Cells, agree in form and in structure of the walls with the woods cells, but they represent earlier stages, and therefore contain more living matter than do the woody fibers. Starch is nearly always present in them, and sometimes chlorophyll and tannin are found. (Fig. 44.)

CHAPTER VIII.
TISSUES OF NUTRITION.

1. ABSORPTION SYSTEM. This consists of tissues whose function it is to absorb from the earth the water containing in solution the various mineral salts that will be of benefit to the plant, and which, with the food received from the air, constitute its entire nourishment. The part of the higher plants devoted to this work is the root hair, which has had a gradual development from the lowest algæ to its present structure. The root hairs are true trichomes of the epidermis of the root, and have very thin walls, through which osmosis can readily take place. (Fig. 45.)

2. ASSIMILATION SYSTEM. The main tissues of assimilation are those in the leaf or leaf-like organs, and the important and active agents are those cells that possess chlorophyll. The leaf consists of a stalk, the blade, and, in some cases, two stipules. The petiole carries the vessels from the stem into the leaf, and has various supporting elements, according to the needs of the leaf. The blade consists of these vessels and some of their protecting elements which are collectively called the ribs or veins and veinlets. These ramify either in more or less parallel lines, as in the Monocotyledons, or irregularly in a reticulat-

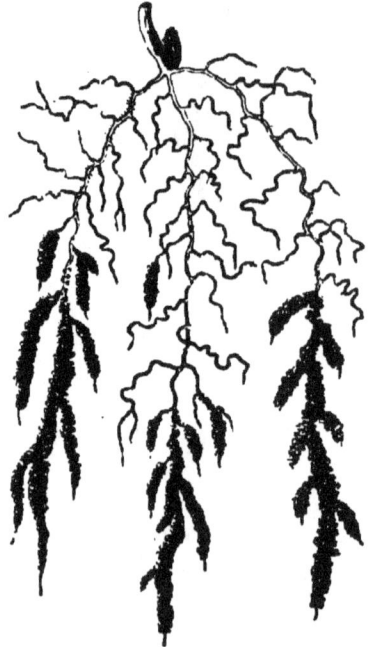

Fig. 45.—Young roots of Triticum vulgare with root hairs and attached earth. (Tschirch).

The Mesophyll alone will here be described, leaving the veins to be discussed under Conduction System and the Stomata with the Respiratory tissues.

The Mesophyll consists of thin-walled parenchymatic cells, which are generally loosely arranged, giving to it the name of Spongy Tissue. In some leaves there is a single or double row of cells, which are somewhat cylindrical or brick-shaped and which are placed with their longest diameters at right angles to the epidermis. They are thin walled and filled with chlorophyll. They are called Palisade Cells, and by means of their chlorophyll assimilation of the carbon dioxide of the air is carried on.

According to the arrangement of the palisade cells three types of leaves have been described.

1. Leaves without any palisade cells. Centric.

2. Leaves with palisade cells on both surfaces. Iso-lateral.

3. Leaves with palisade cells on the upper side only. Bi-facial.

To the first belong many herbs which grow in the shade, also many stipules and calyx lobes, the latter having little or no functions of assimilation.

In the Iso-lateral type a single or double layer of palisade cells is found, with a spongy mass of parenchymatic cells between forming a middle layer. In general leaves that are affected on both sides equally by the sun have this internal arrangement, hence it is found more often in erect monocotyledonous leaves, as in Aloes. It is also found in Senna, Eucalyptus and Lactuca. (Fig. 48.)

The third type, Bi-facial, is the more common. It is found in most leaves which are horizontally placed with reference to the sun's rays. Jaborandi, Absinthium, Conium, Mentha piperita, Laurel, Matico, Uva Ursi, etc., in which the palisade cells may be one, two or five rows in thickness.

It is an easy matter to determine most of the official leaves by means of their external shape and internal arrangement, but in a powdered condition the determination becomes very difficult, and the epidermal cells, the trichomes and

Fig. 46.—Leaf of Digitalis purpurea showing veins or conducting vessels. (Planchon).

ed fashion, as in the Dicotyledons. (Fig. 46.) Between these fibro-vascular bundles the fundamental tissue of the leaf the parenchyma or mesophyll, is arranged in various ways. Surrounding the whole surface is the epidermis, which is a direct continuation of the epidermis of the stem.

Thus, each leaf has portions of at least three distinct types of tissue: The Epidermal tissues, with its modifications, stomata and trichomes; the Conducting tissues, with, perhaps, some mechanical support; veins, and the spongy Internal tissue of the leaf proper, namely, the Mesophyll, which makes up the tissues of Assimilation. (Fig. 47.)

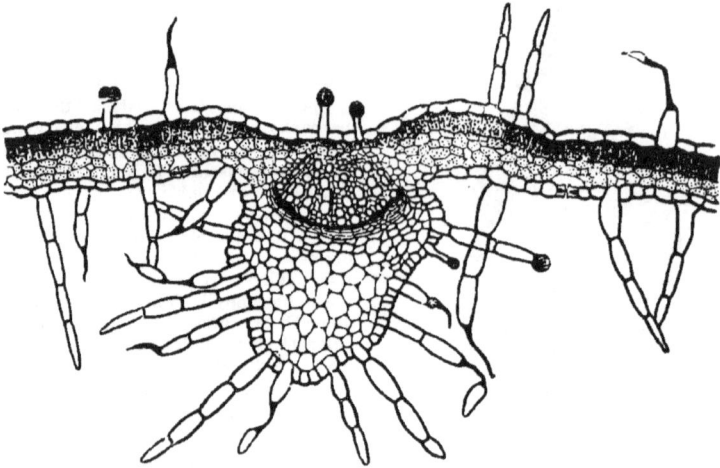

Fig. 47.—Cross section of leaf of Digitalis purpurea, showing finer anatomy of conducting vessels, also hairs and glands. (Tschirch).

the presence or absence of the palisade cells are of great importance. (a)

3. THE RESPIRATORY TISSUES are those which enable the plant to take in food from the atmosphere in the form of carbon dioxide and to give off oxygen and watery vapor. They are the tissues of gaseous exchange. They occur in the epidermis of leaves as the Stomata and as Water Pores, and in the epidermis

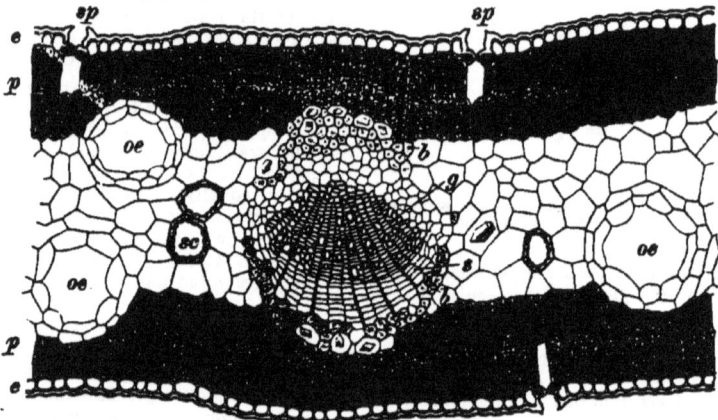

Fig. 48.—Cross section of Melaleuca minor showing *e*. Epidermis, *p*. Palisade cells, *sp*. Stomata, *sc*. stone cells, *oe*. oil reservoirs, (Ol. Cajeput), *g*. vessel portion (xylem)—*s*. sieve portion (Phloem), with *b*. bast fibres of the same. (Tschirch).

of stems as Stomata and Lenticels. As a part of the respiratory tissues Intercellular Spaces should be borne in mind, as they provide means of communication between the internal tissues and the stomata and lenticels.

The Stomata are peculiar modifications

(a) Meyer, Officielle Blätte und Krautter, Halle, 1882. Lemaire, Determination histologique des feuilles medicinales. Paris, 1882. Vogl, Atlas der Pharmacognosie. Tschirch und Oerstele, Atlas der Pharmacognosie. Moeller, Microskopie.

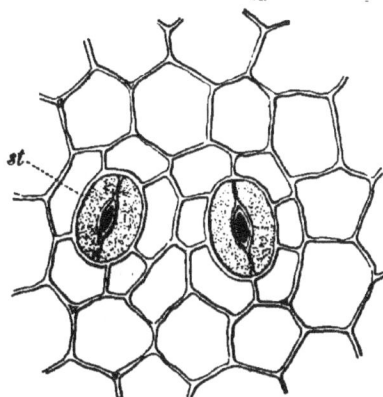

Fig. 49.—Leaves of Uva Ursi, under surface showing *st.* stomata, chink and guard cells. (Vogl).

of the epidermis. In the higher plants they are composed of two guard cells,

as practical guides in the determination in the leaves of the official plants(a).

The number of stomata varies widely, from 0 to 40 to 300 on an average to the square mm. Some leaves have as many as 700. It has already been said that the number on the upper and lower sides may be unequal, the majority of leaves having more stomata on the under side.

The function of the stomata is the interchange of gases and the evaporation of water, the activity of these functions depending upon the opening and closing action of the guard cells, which have a special action. When the guard cells are filled with water, they become turgid, thereby making the opening larger; active evaporation can then take place until the turgor of the cells is gradually diminished, when the opening becomes smaller

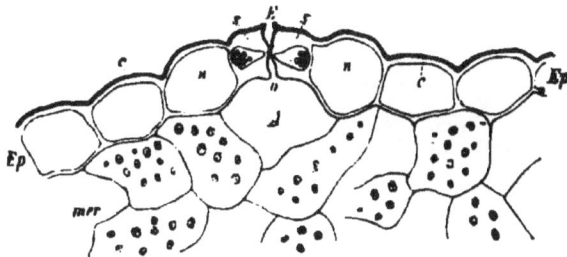

Fig. 50.—Cross section of leaf of Mentha piperita showing *a.* stoma: *E.* opening, with front and hind chambers, *o.* opening, *t.* breathing space. *A. ss.* guard cells with protoplasm. *Ep.* epidermis, *c.* cutinized and thickened outer wall. (Tschirch).

generally oval in shape from a surface view, between which a space is left for the passage of gases into a breathing cavity beneath. (Fig. 49.)

On cross section they are more complicated; in the figure (Fig. 50) ss. are the two guard cells, between which lies the opening E. and underneath the breathing space A. On looking at the cross section the external cutinized layer of the epidermis may be seen. This varies in thickness in every plant. Beneath this in the chink of the opening lie two chambers, the front and back, which vary greatly in the different stomata.

Stomata sometimes lie superficially, at other times, though less commonly, they lie immersed in the tissues of the leaf. These points of variation in shape and size, and degree of cuticularization, serve

and the stomata are thus automatically closed. (b)

In the determination of powdered drugs the stomata are of importance.

Water Pores are not properly respiratory tissues. They are similar in shape to the stomata, but differ in the immobility of the guard cells and in the fact that they generally lie at the end of a fibro-vascular bundle of a leaf, and hence are found in greater number along the edges of the leaves.

Lenticels are respiratory tissues occurring in the stems where the stomata of the epidermis have been lost by the process of secondary growth already described. They occur on stems, leaves

(a) Tschirch, Angewandte Pflanzen Anatomie, p. 434.

(b) Schwendener, Berichte d. Berliner Akademie, 1881.

and on roots in the shape of small rounded or oval or warty elevations of loose and irregular corky tissues; this looseness of arrangement of the cells allows of the free exchange of gases to the stem that otherwise could not occur.

Intercellular Spaces. These exist between the cells of plants partly in order to allow the gases taken in by the stomata and lenticels to go to all parts of the plant, and conversely to allow the products of oxidation and other products of gaseous interchanges to escape. Such spaces are found in most tissues of the plant, save the epidermis. They vary greatly in shape and size, and are often filled with products of secretion and excretion, to be later discussed. They frequently serve as carriers of water and are quite common in plants growing in marshy grounds. (Fig. 51.)

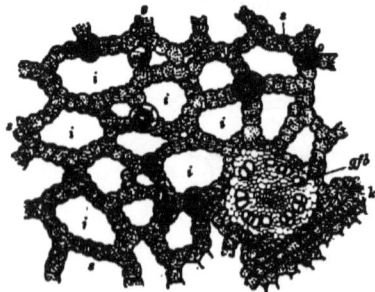

Fig. 51.—Cross section of Rhizome Calamus showing, i, intercellular spaces, s. starch conducting parenchyma, o. oil cells, G. f. concentric vessel bundle with xylem surrounding phloem. (Tschirch.

4. CONDUCTING SYSTEM.

In the lowest plants there is no conducting system proper. The cells lying either singly, in chains or surfaces, absorb freely the necessary food materials, each for itself. In the higher Thallophytes, the seaweeds and mushrooms are met modifications which foreshadow a conducting system in plants; but only in the Bryophytes, in the mosses proper, is there to be found a true series of elongated cells set apart to do an especial work in the conduction of food materials. From the mosses on upward the conducting system becomes more complete and more intricate. At first it is a simple string of elongated tubular cells. Later it develops into a system which it is the purpose here to describe.

It can readily be seen that as plants get larger the growing tips, as the leaves and the flowers, are further removed from the source of supplies, i. e., the roots, and there must be established a quick line of communication between the extremities, hence the necessity for a system of cells capable of transporting and communicating fluids, both from the root to the leaves and from the leaves back again to the roots.

The groups of tissues of the conducting system, together with the mechanical fibres that support them, and which have been already described are called the Fibro-Vascular Bundles.

As the elements of the Fibro-Vascular Bundle have all been developed from the same primary meristematic tissue, it is but reasonable to remember that all degrees of variation will be found in a careful study of the elements of such a bundle. For convenience, however, the typical elements have been distinguished from one another, and named.

Naegli has divided the Fibro-Vascular Bundle into two parts, from a purely anatomical standpoint. One he calls the Xylem, or the woody portion, and the other the Phloem, or the bast portion. According to his classification the Xylem contains

Wood fibres,
Vessels,
Parenchyma,
The Phloem contains
Bast fibres,
Sieve tubes,
Parenchyma.

Haberlandt has proposed a classification which divides the bundle into parts more in accordance with the physiological functions of the plant, as follows:

1. Vessel portion—Hadrom.
   (a) Vessels,
   (b) Tracheids,
conducting water,
   (c) Woody parenchyma, conducting water and plastic materials.
2. Sieve portion—Leptom.
   (a) Sieve tubes and accompanying cells,
   (b) Cambiform cells,
       conducting plastic materials.
   (c) Accompanying cells (Geleitzellen).

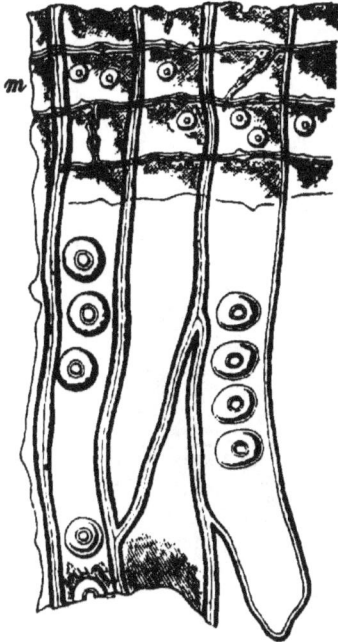

Fig. 52.—Tracheids of Pine wood radial long section with border pores. *m.* medullary rays. (Tschirch).

3. Parenchyma sheath, conducting soluble plastic materials.

In this classification it will be noticed that the bundle is considered apart from its mechanical supports, and that the terms Hadrom and Leptom are synonymous with Xylem and Phloem, from which the wood fibres and bast fibres have been omitted.

Much variation will be found in individual bundles, in the number and kinds of elements present, some bundles containing only one or two of the elements. Such variations should be borne in mind in the practical study of tissues. In general, however, anatomists recognize as component parts of the conducting system the following: The vessels, tracheids, and conducting parenchyma in the Hadrom; the Sieve tubes, Accompanying cells (Geleitzellen), and Parenchym in the Leptom.

1. The Vessels form the most conspicuous portion of the internal structure of the plant. Their function is to carry water with inorganic salts in solution up from the root hairs to mingle it with the plastic materials that are to be brought down from the leaves. The vessels are of two main types, Ducts and

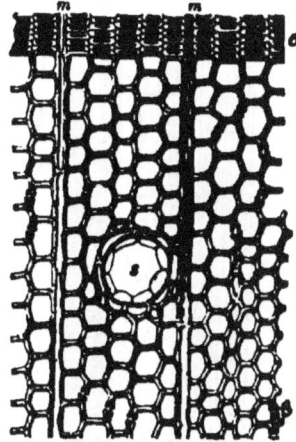

Fig. 52. *a.*—Tracheids of Pine wood on cross section. *m.* medullary rays; *c,* fall wood. *s.* resin passage in spring wood—border pores seen in cross section *b.* (R. Hartig).

Tracheids, both being by some authors included under the name of Trachea.

The Tracheids are of a more elementary type of cell and make their appearance lower down in the vegetable world than do the Ducts. In the lower plants they are simply elongated cells with a well-marked lumen and oblique cross walls. They first assume their charac-

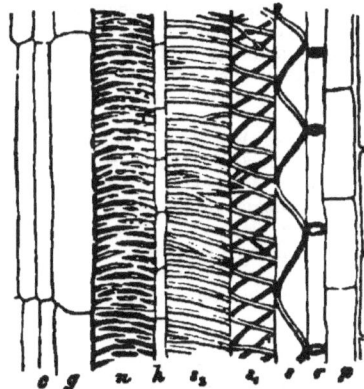

Fig. 53.—Types of ducts. *r.* annular. *s. s., s,,.* spiral ducts, with simple and double spirals, *n.* reticulated.

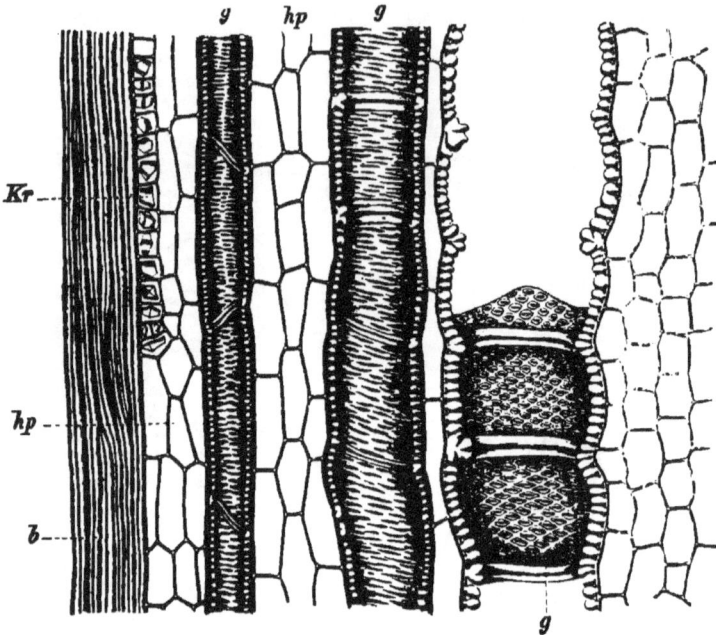

Fig. 54.—Longitudinal section of Glyoyrrhiza glabra showing, *g.* vessel with pitted and border pore markings in the walls, *hp.* wood parenchyma, *b.* wood fibres, libriform. (Tschirch).

teristic shape in the Ferns. Their length varies. In pines it is 4 mm. The ends of the Tracheids are pointed, and sheath into one another. The walls are generally thickened and the lumen of the cell is generally extensive. The thickening of the walls is due to lignification which takes place irregularly, sometimes forming spiral markings and sometimes thickening the whole wall save in certain spots where border pores may occur. (Fig. 52.)

Simple slit-like pores, such as are found in Libriform tissue, are generally wanting in tracheids. This fact is of practical diagnostic importance.

In some plants the water-conducting tissues consist entirely of tracheids. This is the case in the pines, in the root of Ipecac, and the nerve endings of leaves. The sides of tracheids which border on ducts or libriform tissue are apt to conform in their appearance to those contiguous tissues, therefore sharp distinctions between them cannot always be made.

Ducts. Under this head are included cells which resemble tracheids in many respects, but differ in the fact that their end walls are absorbed, thus forming long open tubes which may run the entire length of the plant without any cross walls. Although Ducts were originally closed cells, by the absorption of their cross walls they cease to be cells in one sense of the word.

The modification which Ducts have undergone, through thickening of the cell wall by lignification, are so marked that special names have been given to certain characteristic forms. The principal types are Annular, Spiral, Reticulated and Pitted Ducts, with intermediate forms between. (Fig. 53.)

An Annular duct (r.) is one in which the thickening of the cell wall takes place in the form of a ring, making a series of bands around the inside of the tubular vessel. This form occurs in the stem of Conium, and Radix Podophyllum.

Spiral ducts have the lignification dis-

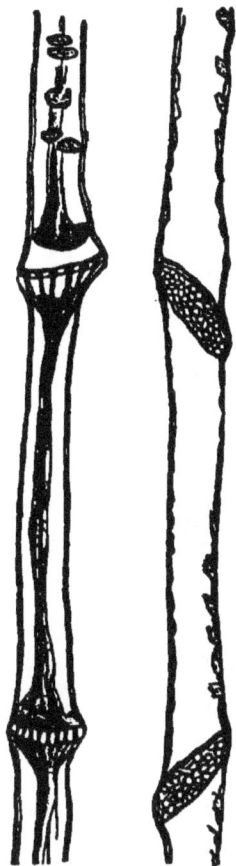

Fig. 55—Sieve tubes from Poppy capsule. (Dippel).

posed in the form of spirals. (S,S₁,S₂.) There may be from one to several on the inner wall of a single duct. The spirals generally coil from right to left, but in some cases two or more will coil in opposite directions, forming a double set of spirals. Some variation of this form is found in the corollas of most flowers, in the stem of Conium, in Bulbus Scillæ, etc.

Reticulated ducts have net-shaped thickenings on the inner walls. (n.) Sometimes they are quite regular, forming ladder-like markings, and again they are extremely irregular.

Pitted ducts are quite different from the preceding forms in that their walls are wholly thickened but densely per-forated with pores, either bordered or simple, or both, and in form round, oval or slit-shaped. Typical pitted ducts may be seen in Radix Glycyrrhiza, Tuber Jalap. (Fig. 54.)

Primary vessels are generally annular or spiral, presenting a greater unthick-ened interior surface, but the vessels of secondary growth are generally reticulated or pitted.

Parenchyma. Numerous parenchy-matic cells may be found interspersed among the elements of the Xylem that serve for purposes of water conduction. As a rule, they offer little in the way of diagnostic characteristics.

2. Leptom, or the Conductors of Plastic Materials. The leaves and young grow-ing stems contain plastic materials which, when combined with the water and inorganic salts brought by the ducts and tracheids from the roots, serve the purposes of anabolism. The Leptom is a series of tissues (according to Haber-landt, the conducting elements of the Phloem), whose special function it is to convey these worked-over materials down from the leaves, where the active transformations take place, to those parts of the plant where they are needed.

The Leptom consists of Sieve Tubes, Cambiform Cells and· Accompanying Cells, in fact, all the elements of the Phloem, except the bast fibers, which be-long to the protection system.

Lactiferous Tissues, or Milk Tubes, are in part plasma-conducting vessels and in part secretory organs.

(a) Sieve Tubes are long cells whose end walls are perforated, forming a sieve-like plate through which the plastic materials can pass. These sieve plates may be horizontally arranged if the ends of the cells were originally straight, but they are generally much inclined. As a rule, the sieve tubes in primary tissues are more apt to be horizontal, while those in tissues of secondary growth are apt to be inclined.

The walls of sieve tubes are generally thin and composed of cellulose. The sieve tubes vary greatly in size, but as a rule they are smaller than the vessels. They are often very irregular in form, being large and swollen at the points where the sieves occur and much con-tracted between. (Fig. 55.)

Their contents are plastic, consisting

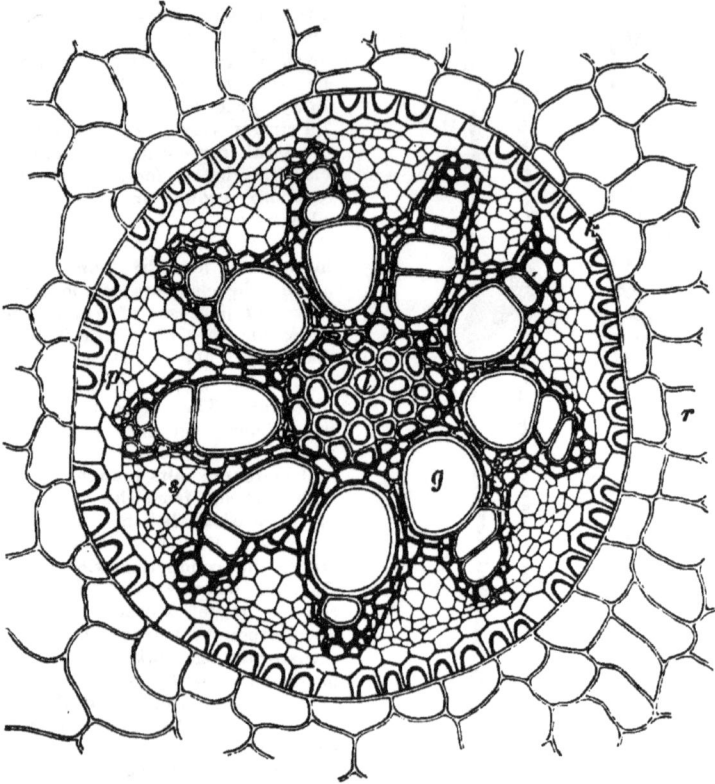

Fig. 56.—Cross section of *Radial bundle* in root of Veratum viride. *k.* endodermis sheath, *g.* vessel portions, *s.* sieve portions. (Tschirch).

largely of Albuminoids. These often form by deposit upon the upper surface of the sieve a callous growth which is apt to appear in autumn and winter, and to stop up the openings of the sieve. This may be redissolved in the spring, or, in some cases, it may continue until it obliterates the sieve tubes entirely, as is the case in Radix Glycyrrhiza.

(b) The Cambiform Cells are exceedingly thin-walled, elongated cells with sharpened ends. They closely resemble Cambium cell, and are filled with protoplasmic contents. Their walls have no traces of pores.

(c) Accompanying Cells (Geleit Zellen) are cells accompanying the sieve tubes. They are seen to best advantage on the cross section of the bundle, where they appear more or less quadrangular, with small lumen and very thin walls. They transmit protoplasmic substances and are connected ofttimes by pores with the sieve tubes.

In specimens of dried drugs it is almost impossible to distinguish the accompanying cells or the cambiform cells, as they are so delicate.

3. Conducting Parenchyma. As in the Hadrom parenchymatic cells were found taking on the function of conducting water, so in the Leptom portion of the bundle there are numerous parenchymatic cells which conduct plastic substances. The prominently marked starch sheath of many Monocotyledons and young Dicotyledons belongs to this group of tissues. The parenchymatic tissues of the phloem are important agents in carrying on this function.

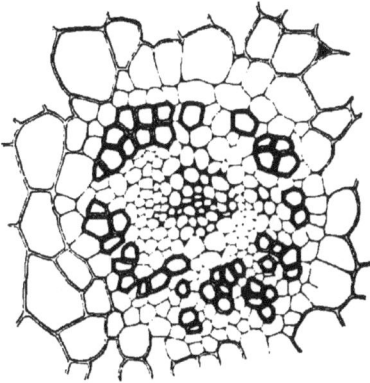

Fig. 57.—Cross section of *Concentric bundle* of Rhizome of Iris with central sieve portion surrounded by vessel portion. (Tschirch).

Types of the Fibro Vascular Bundle.

The individual tissues which collectively form a Fibro-vascular bundle may be grouped in various combinations. Anatomists describe three principal types based upon the relationship that the vessel portion, or Xylem (in the wider sense), bears to the Sieve portion or Phloem (in wider sense). These are

1. Radial.
2. Concentric.
3. Collateral, of which three forms are described.

1. The Radial Bundle is one in which the xylem and the phloem lie side by side, alternating with each other radially. The xylem starts at the center and builds strong ribs of tissue between which lie the phloem elements. Radial bundles are characteristic of young roots, and ac-

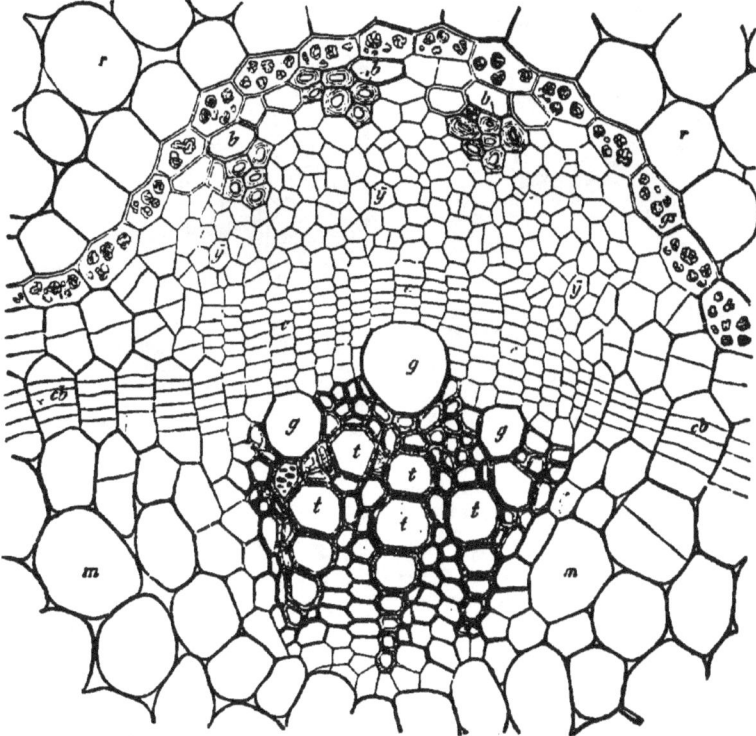

Fig. 58.— Cross section of developing open Collateral bundle, *g.* vessels. *t.* tracheids, with libriform between, constituting the xylem part of the bundle *c.* cambium between it and, *y.* sieve tubes, *b.* bast fibres, with cambiform and and accompanying cells between, constituting the phloem part of the bundle. (Sachs).

cording to the number of the bundles the roots are known as diarch, triarch, tetrarch, etc. Few of the official roots have radial bundles. Sarsaparilla, Aristolochia. (Fig. 56.)

2. The **Concentric Bundle** is one in which one of the elements encircles the other more or less completely. In some cases the Xylem is in the centre and surrounded by the phloem, (Aspidium). In others the phloem is in the centre and is surrounded by the Xylem. (Calamus, Iris, etc.) (Fig. 57.)

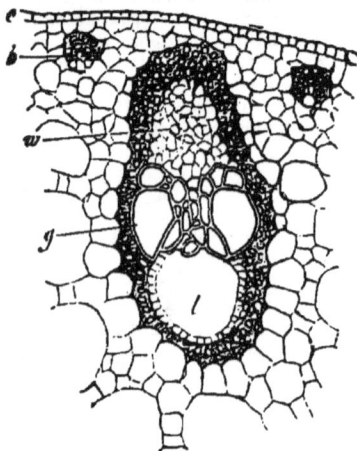

Fig. 59.—*Closed Collateral* bundle from the leaf of Calamus, *g.* vessel portion, *w.* sieve portion. (de Bary).

Concentric bundles are more common in Monocotyledons, where they form strings of tissue irregularly disposed in the parenchymatic tissues.

3. The **Collateral Bundle** is characterized by the Xylem and phloem lying side by side in a tangential direction, the xylem being towards the centre. This is the type of bundle of most Phanerogamic stems and roots. Three forms of collateral bundle have been described.

　　a. The Open Collateral.
　　b. The Closed Collateral.
　　c. The Bi-collateral.

a. The **Open Collateral** bundle is characteristic of Dicotyledons. The centre of the root or stem and generally the larger part of it is occupied by an indefinite number of Xylem elements radiating from the axis, forming a solid cylinder. This is surrounded by a thin layer of

growing tissue, the Cambium, which separates it from a zone of combined phloem

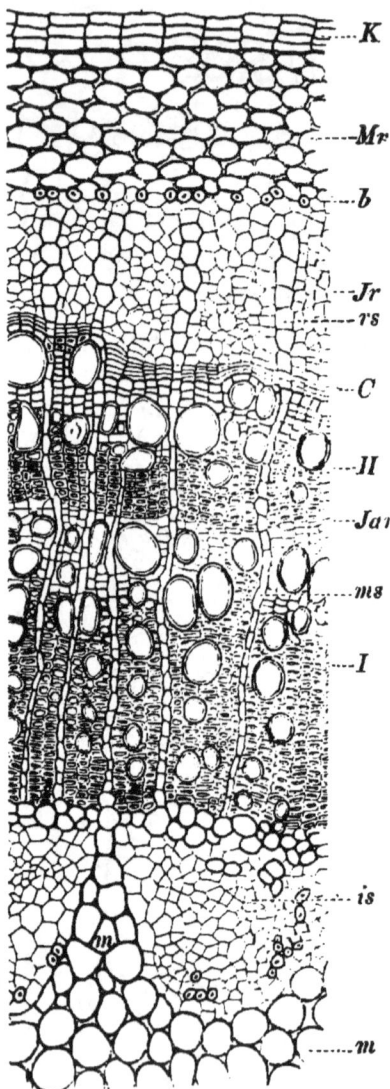

Fig. 60.—Cross section of two year old stem of Solanum Dulcamara showing *Bi-Collateral* bundles, *is* inner sieve portion, *Jr.* outer sieve portion in secondary bark with cambium and vessel portion between, *rs, ms.* medullary rays, in sieve portion and vessel portion, respectively. (Tschirch).

elements. In young roots and stems the individual bundles lie distinct and separate. As the growth of the root or stem proceeds they gradually coalesce, the cambium zone becomes complete and the bundles remain separated only by thin radiating lines of tissue termed the Medullary Rays. By virtue of the meristematic power of the Cambium, the Xylem is constantly growing centripetally, while the Phloem grows centrifugally.

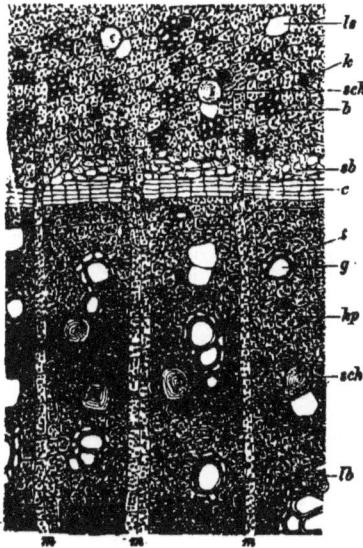

Fig. 61.—Cross section of root of Althæa officinalis. *c.* cambium; *m.* medullary rays, *sch.* mucilage cells, *b.* bast fibres, *hp.* woody parenchyma filled with starch, *sb.* sieve portions, *lb.* libriform. (Tschirch).

The bundles are therefore capable of indefinite extension, and are consequently called Open Collateral bundles.

b. The Closed Collateral bundle is characteristic of the Monocotyledons. The elements are grouped together without any cambium, and are generally surrounded as a whole by a circle of schlerenchymatic fibres which checks the lateral growth of the bundle. To this arrangement is due the fact that Monocotyledons after forming their bundles do not grow in thickness but only in height. (Fig. 59.)

c. The Bi-Collateral Bundle has two strands of Phloem, one on each side of the Xylem. It is found in the Cucurbitaceæ and Solanaceæ, etc. (Fig. 60.) Solanum Dulcamara.

Secondary Growth is the term given to the process which takes place in the stems of most Dicotyledons, increasing their size and giving rise to a set of new tissues known as Secondary Tissues. These as individuals have already been described in the elements of the xylem and the phloem of the fibro-vascular bundle, and it here remains to look at the process in general and to describe a few tissues not before spoken of.

In Monocotyledons there is as a rule no secondary growth, a few exceptions occurring in the Lily family, (Dracæna, Aloes, Smilax), and, in consequence, the Monocotyledons generally possess long thin stems which increase but little in diameter. In Dicotyledons, on the contrary, there is a regular growth. New tissues are constantly being formed in annular layers by means of the Cambium. The Cambium is a meristematic tissue having its origin in the primary meristem of the plerome cylinder, and by its divisions new elements are added to the phloem and the xylem. The cambium cells are elongated and have their walls at right angles. They are somewhat brick-shaped, with their longest diameter vertical. The cambium cells increase by tangential division, at times building up the sieve elements, again forming the woody elements which in general are formed more rapidly.

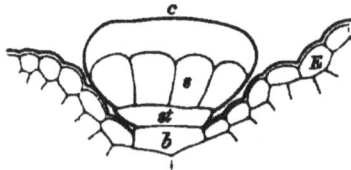

Fig. 62.—Cross section of Oil gland from leaf of Mentha piperita. *b.* base. *st.* supporting cell, *s.* secreting cells, *c.* cuticle, *E.* epidermis of leaf. (Tschirch).

Medullary Rays. Between the fibro-vascular bundles of most Dicotyledons and Gymnosperms thin strips of parenchymatic cell are found. These are the Medullary Rays, connecting the pith with the epidermis. Two kinds are to be distinguished:—Primary medullary rays which represent the early divisions between the bundles, and Secondary medullary rays, which occur in the process of secondary growth.

In shape the cells of the medullary rays are iso-diametric with slightly thickened right-angled walls, which are provided with simple pores. Their function is to convey nourishment to different parts of the plant during the growing time; and during the resting period of the plant they serve as storehouses of reserve material, starch, etc. The number and size of the medullary rays, the shape and size of the markings of the individual cells, are of diagnostic importance for pharmacognostical purposes.(a)

Woody Parenchyma is also a product of the cambium by means of crosswalled division. They are iso-diametric in cross section and somewhat elongated. They conduct plastic materials by means of their thin walls, which are provided with round or oval or simple pores. During the resting period of the plant these cells contain reserve materials, mainly starch.

5. SECRETORY AND EXCRETORY SYSTEMS. In the process of plant developement many compounds are produced which the plant either stores up for future use or gets rid of entirely. Many of these secretions have especial reservoirs adapted for their reception, while others use ordinary conducting cells.

A large number of these secretions have no economic value in contributing to the growth of the plant, but may be

(a) Nordlinger, Vergleichende Anatomie der Wichtigste deutschen Wald und Garten holz arten, Stuttgart. 1881.

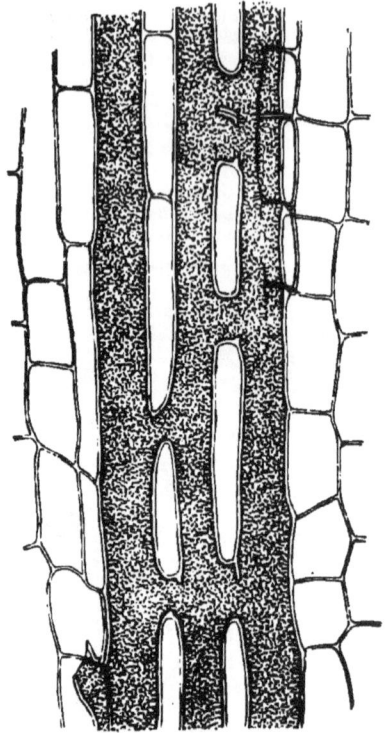

Fig. 64.—Branched milk tubes in the cortex of stem of Lactuca virosa, which contain Lactucarium. (de Bary).

of use in attracting insects, healing wounds, or in preventing the ravages of animals. The secretion of oil and its related products may have its influence upon the temperature of plants by preserving them from excessive heat in the

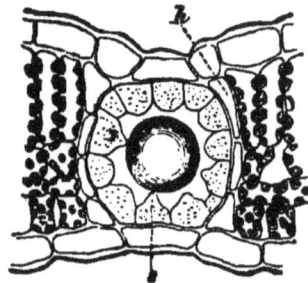

Fig. 63.—Cross section of resin passage in leaf of Pinus; oil cells in leaf.

day time and cold at night. (Haberlandt, p. 325.)

Haberlandt classes the organs of this system as follows:

(a) Secretory organs, glands, external and internal channelled organs for resin, oil, mucilage and gum.

(b) Excretory organs, mucilage, resin and oil holders, tannin and crystal sacs. Intercellular excretory reservoirs.

Glands. These may be external or internal. The external glands are generally at the extremity of epidermal hairs, which at times are quite short. They may be one celled or composed of many cells. Mint. (Fig. 51.) By the gradual filling up of the secretory cells these glands grow larger, forming distinct sacs, filled with the secretion. The ethereal oils are generally found in such reservoirs.

Official plants bearing external glands are Hyoscyamus, Tabacum, Mentha, Thymus, Lavandula, Humulus, Matricaria, Achillea, Arnica, Absinthium, Kamala, etc.

Internal Glands and Excretory Cells are the names given to a number of cells varying in shape and size which are generally contained within the parenchymatic tissues of the plant organs. They are generally large sac-shaped cavities filled with secretions, oils, mucilage, etc., and are generally designated according to the character of their contents. The most prominent are those containing resin, oil and crystals. Such secretory organs are to be found in Rhizomes, Calamus, Zingiber, Galangal, Curcuma, Zedoary, Cortex Cascarilla, Angustura, Radix Valerian, Folia Laurus, Sassafras, Matico, Fructus Cubeba, Cardamomum, Sinapis, etc. (Fig. 63.)

Channelled Secretory Organs. These are generally formed in the intercellular spaces of the plant. They are much elongated and are surrounded by secretory cells. Conifers, Umbelliferæ.

Secretory Vessels are formed in the tissues of the plant by the coalescence of a number of secretory cells, thus making irregular tubes which may be branched or unbranched. These generally contain the so-called milky juices, and are met with in the Euphorbiaceæ, Asclepiadaceæ, Campanulaceæ, Lobeliaceæ, Papaveraceæ and Apocynaceæ. (Fig. 64.)

Crystal Cells are generally thin-walled and sometimes cutinized or suberized. They vary according to the shape of the contained crystals, being rather regular in those containing rhapides.

Tannin occupies the typical parenchymatic cells of the plant, which, when so employed, are generally suberized and have a brownish or reddish appearance.

The study of those portions of the plant devoted to excretion and secretion are as yet little understood, and the student is referred to the bibliographical references for a more complete account of them.

CHAPTER IX.
REPRODUCTIVE TISSUES.

There are few types of tissue that have not at least been touched upon in the foregoing pages, and although there are no essential differences in the tissues of reproduction, it is, nevertheless, of advantage to recapitulate somewhat concerning them. The organs of reproduction are the pistils and the stamens, and each have some elements that are of use in a practical determination of many drugs, such as the official herbs, flowers, fruits and seeds. In herbs and flowers the pollen grains are the only elements that have not received some attention; they will therefore be treated somewhat hastily in this place.

Pollen Grains originate by cell division in the anther sacs. When mature they vary greatly in shape, size, color and markings. Specimens of drugs that are to be examined for pollen grains should be placed in some oily substance, as this preserves the markings of the grains better than water. Specimens may also be examined in water, as many of the older figures of pollen grain were made from grains that had been examined in that medium. The differences in the shapes and other peculiarities of the grains of the official herbs and flowers it is impossible to treat of in this place, but the student is called upon to bear their importance in mind, and is referred a treatise on "Pollen" by M. Pakenham Edgeworth, London, 1877, where the literature of the subject up to that time is given.

The parts of the pistil have already been considered. In the study of the mature pistil, in particular the seeds, a great deal could be written of the numerous modifications that take place in the cells of these parts. In general they belong to the parenchymatic type, while the chief interest in them is the great amount of thickening and distortion that takes place. The changes that take place during fertilization are of interest from a scientific botanical standpoint rather than from a practical analysis point of view.

CHAPTER X.

MICRO-CHEMICAL REACTIONS.

It is far from the purpose of the present chapter to do any more than to point out in as brief a manner as possible some of the more important micro-chemical reactions of plant structures or contents; the student who would follow them out to any further detail is referred to "Botanical Micro-technique," by Zimmerman, H. Holt & Co., and Hervey's translation of the "Microscope in Botany," by Behrens. In these volumes a full consideration of the subject will be found, with copious bibliographical references.

Tests for Cell Walls. It has already been pointed out that the cell wall may undergo at least five modifications from its chemical basis of cellulose. Each of these has some micro-chemical tests that are of value.

Cellulose. Strong sulphuric acid causes cellulose to swell, turn blue and dissolve. Schweizer's reagent, cupra-ammonia, dissolves cellulose completely. It should be made fresh, as it does not keep well. Iodine and sulphuric acid give a blue color. Chlor-iodide of zinc, blue to violet coloration. Hematoxylin in weak solution stains cellulose wall blue to black, depending upon the time of staining. I prefer Delafield's solution for this purpose.

Lignification. Lignified cell walls are insoluble in Cupra-ammonia, and they color yellowish to brownish by Chlor-iodide of Zinc or Iodine and Sulphuric acid. Several aromatic compounds, as Phenol, Thymol, Orcein, Aniline, Phloroglucin, etc., give characteristic reactions. (See table Zimmerman, p. 141.) Phloroglucin in water or alcoholic solution, preceded or followed by dilute Hydrochloric acid, gives a beautiful cherry-red coloration to lignified cell membranes. Fuchsin stains lignified cell walls a deep and persistent red.

Cutinized and Suberized Cell Walls have about the same micro-chemical properties. They are insoluble in sulphuric acid, are colored yellow or brown by chlor-iodide of zinc, and iodine and sulphuric acid. Concentrated caustic potash causes a yellowish coloration which becomes more intense on heating.

Mucilaginous Modification. This can be readily detected by the extreme degree of swelling that takes place on the addition of water. The various gums have a number of different reactions which must be sought for in the works of reference already quoted.

Tests for Cell Contents.

Nitrogenous Contents: Protoplasm. It has already been pointed out that the albumenoids of the plant are numerous and very complex, and a great many characteristic reactions have been found for different members of this general group. Only those general reactions will be here mentioned that have been in long use. With iodine the proteids take on a yellowish or brownish or even black coloration, according to the strength of the solution. Concentrated nitric acid gives a yellow color, the Xantho-proteic reaction, and Millon's reagent (a mixture of mercuric and mercurous nitrate with nitrous acid, prepared by dissolving one part of mercury in two parts of nitric acid and then diluting with twice the volume of water) colors proteids red in various shades; slight warming brings out the reaction more rapidly.

Aleurone grains. In searching for aleurone grains the previous treatment of the sections is of importance. They should not touch water in any form, as aleurone is soluble in that medium. The material to be investigated should first be fixed, preferably with absolute alcohol or picric acid alcohol, and then treated according to the desire of the investigator to bring out the inclusions or the fundamental mass or ground substance of the grains. To bring out the ground mass, staining in an alcoholic solution of

Eosin gives a reddish coloration, and the crystalloids remain yellowish if picric acid was used to fix. To bring out the crystalloids we prefer to use acid fuchsine. This stains the crystalloids an intense red. Permanent preparation should be preserved in balsam, glycerine or some oil.

Non-nitrogenous Contents. Of the non-nitrogenous cell contents there are to be considered, with a greater or less degree of minuteness, oils, fats, starch, sugar, amylodextrin and sulphur.

The Fats and Oils can generally be detected as small globules of a yellowish or brownish color, circular or slightly oval and of a high refractive index. They are generally found lying free in the cell cavities. In powdered drugs they are dissociated and sometimes run together in masses. Micro-chemically, osmic acid, 10 per cent., stains them black; they are saponified by the alkalies, a weak solution of KOH being sufficient; tincture of alcanna stains fat and oil globules bright red. Absolute alcohol distinguishes the fixed oils and fats from the essential oils; the former are insoluble, the latter soluble. A characteristic reaction is seen when a section of a plant to be investigated is placed upon the slide in glycerine and slightly warmed. The fats and fatty acids melt, at first in drops, and then by slow cooling crystallize in long needle-shaped crystals collected in bundles.

Starch is distinguished both morphologically and micro-chemically with great readiness. It makes a characteristic reaction with iodine solution, varying with the strength of the solution and the length of application from blue to violet to black; alkalies destroy the coloration and acids restore it.

The Constituents of the Cell Sap are the most numerous of the cell contents, and much research is necessary to differentiate all the substances thus far described; hence they can be hardly more than touched upon. These materials in solution include hesperidin, inulin, asparagin, tyrosin, aloin, sugar, mucilage, tannins, alkaloids, glucosides, bitter stuffs, ethereal oils, gums, resins, rubbers, milky juices, balsams, plant acids, and various crystals. A number of these are of little importance save in the most detailed investigations, while others are constantly determined by chemical means, and their presence can frequently be made evident by means of the microscope. The more important constituents from our present standpoint are tannin, resins, sugar, wax, the alkaloids, and the glucosides.

Sugar. Although Trommer's and Fehling's tests are most often given for the detection of sugar, it is not an easy matter to determine its presence in small quantities; Phenylhydrazine acetate has given satisfactory results in our experience when the amount of sugar has been very small.

Tannin is found extensively in plants, generally dissolved in the cell sap, especially of the bark; it is often found in granular form also. As a general micro-chemical reagent for tannin, ferric chloride is used, either in aqueous or alcoholic solutions. Ferric sulphate and acetate give less intense reactions. If a substance is impregnated with any of these mixtures, the masses of tannin or the cellulose membranes impregnated with tannin give a bluish or greenish reaction, strong solutions giving a blue-black or greenish-black reaction. The tannin in galls, oak for instance, give the bluish reactions, whereas rhatany, coffee, and male fern give the greenish reaction. Potassium bichromate is also of value, giving after some time a reddish coloration. A dilute solution of zinc chloride gives a reddish violet coloration with tannin.

Resins are found either fluid or in more or less solid granules, sometimes lying in the cell wall, sometimes in special secretory reservoirs, or partly saturating the cell walls. The resins are generally brownish in color, and if in grains, irregular in shape. Tincture of alcanna, 50 per cent., stains resin a cinnamon red. A solution of equal parts methyl-violet, fuchsin, and alcohol (Hanstein) produces a blue, or clear green, or dirty green color.

The Alkaloids and Glucosides, microscopically, are more of a hope than a reality in practical work, for in their natural condition they are difficult if not impossible to recognize. Micro-chemically, however, much can be done in their determination, but the work is for the advanced student rather than the tyro, and to undertake it one should consult the works already noted.

ERRATA.

Page 114, first column, line 26, from above, insert the words "cover glasses" after the word "preferred."

Page 119, first column, lines 29 and 34, from above, read "2-5 mmm" and "30-200 mmm," instead of "2-5 mm" and "30-200 mm," respectively.

Page 120, second column, line 10, from above, read "Agglomerative," instead of "Agglomeratur."

Page 124, first column, line 21, from below, read "Two main types," instead of "the two main types."

www.ingramcontent.com/pod-product-compliance
Lightning Source LLC
Chambersburg PA
CBHW030556270326
41927CB00007B/945